Renewable Energy Sources for Sustainable Development

About the Authors

Dr. N.S. Rathore, Professor (Renewable Energy) is presently working as Dean, College of Dairy and Food Science Technology, Maharana Pratap University of Agriculture and Technology, Udaipur. He did M.Tech with specialization in Energy Studies from I.I.T., Delhi and Ph.D. from Rajasthan Agricultural University, Bikaner. His scientific interests are in the field of mew and renewable energy sources, ecology and environment and application of energy in agriculture. He has contributed more than 105 technical papers and 32 books on various aspects of Renewable Energy, Environment and Agricultural Engineering. He has undertaken 18 research projects and organized 18 International trainings, Seminars, Workshops and Conferences and 12 ISTE sponsored summer school. There are 95 reports, proceedings and popular articles to his credit. He was honored by The Indian Society of Agricultural Engineers (Appreciation Medal and Outstanding Book Award), Ministry of Environment and Forest (Outstanding Book Award), ISTE (Nanda Memorial Young Scientist Award) and ISAE Team Research Award. He has traveled in various parts of India and abroad including U.S.A., Australia, Thailand, Hongkong etc. for the academic activities.

Er. N.L.Panwar is working as Assistant Professor, Department of Renewable Energy Sources, College of Technology and Engineering, Udaipur. He did M.E. with specialization in Renewable Energy Engineering. He is also working as Research Engineer in the AICRP on Renewable Energy Sources (Thermo Chemical Conversion Technology). His scientific interests are in the field of new and renewable energy sources, energy conservation and application of energy in agro-based industries. He has contributed more than 15 technical papers and 2 books on renewable energy aspect.

RENEWABLE ENERGY SOURCES FOR SUSTAINABLE DEVELOPMENT

N.S. RATHORE
Professor (Renewable Energy) *and*
Dean, College of Dairy and Food Science Technology
Maharana Pratap University of Agriculture & Technology
Udaipur (Rajasthan), India
&
N.L. PANWAR
Assistant Professor
Department of Renewable Energy Sources
College of Technology and Engineering
Maharana Pratap University of Agriculture and Technology
Udaipur (Rajasthan), India

New India Publishing Agency
Pitam Pura, New Delhi- 110 088

Published by
Sumit Pal Jain *for*
New India Publishing Agency
101, Vikas Surya Plaza, CU Block, L.S.C. Mkt.,
Pitam Pura, New Delhi- 110 088, (India)
Phone: 011-27341717, Fax: 011-27341616
E-mail: newindiapublishingagency@gmail.com
Web: www.bookfactoryindia.com

ISBN : 9788189422721

Composed and Designed by NIPA

Preface

Strategies and interventions are needed that promote energy as an engine for equitable economic growth and sustainable development energy use is highly unequal between North and South, rich and poor, men and women. This challenge presents an opportunity: adopting ways and means of producing and using energy that are economically, socially and environmentally sustainable thereby making energy an important tool for sustainable development. Today, two billion people rely on traditional fuels such as wood, dung and agricultural residues to meet their heating and cooking needs, entrenching poverty and limiting opportunities. There are number of energy sources, conventional as well as non-conventional, available in rural areas which include kerosene, diesel, petrol and other petroleum product for meeting rural needs. In addition to this, non-conventional energy sources, mostly, new and renewable energy sources in the form of solar energy, bioenergy, biogas, improved chulha, wind energy are also available in plenty in rural areas.

In order to meet essential rural energy requirements and for fulfilling the norms of energy conservation in present energy quest and for best utilization of available conventional as well as non conventional energy sources, it is essential that sources need to be used in integrated manner. Therefore, Integrated Rural Energy Planning (IREP) is essential for developing country like India. The IREP involves number of input, which are basically derived on present energy consumption pattern and decentralized energy planning. The integrated rural energy planning programme is different from other rural development programmes. It require knowledge of a great number of topics. Keeping this in view, this book is written with the primary objective to present the concept and basis of Integrated Rural Energy Planning.

The content of book can easily be grouped into four major aspects. The first aspect includes fundamental and design of energy system and specially design in Integrated Rural Energy Planning System. This include general characteristics of energy and its role in different essential activities, followed by need for IREP and essential of IREP i..e energy assessment & different approaches for IREP.

The second aspect of this book deals with energy integration techniques, energy forecasting and rural energy modeling with special emphasis on computer modeling techniques. In the third aspect, the application of different new and renewable energy sources including solar energy, bioenergy, wind energy, biogas, improved cookstoves. *etc.*

The fourth section of this book consists of techno economic analysis of different energy options and energy. Much of information presented in this book is a basis to acquire an understanding of the Integrated Rural Energy Planning programming, its design, development, implementation, monitoring and feedback evaluation.

Inspite of our best efforts, some errors might have occured in the text. We welcome the suggestions and comments if any, from our readers for further improvement of the book.

Authors are thankful to the staff of Department of Renewable Energy Sources, CTAE for their support and contribution.

Rathore & Panwar

June, 2007 Udaipur

Contents

1

Introduction

1.1 General

Access to energy is a crucial enabling condition for achieving sustainable development. Prudent energy policies and research can play an important role in steering both industrialized and developing countries onto more sustainable energy development paths. Specifically, they can strengthen the three pillars of sustainable development: the economy, by boosting productivity, social welfare, by improving living standards and enhancing safety and security, and the environment, by reducing indoor and outdoor pollution and remediating environmental degradation.

1.2 Energy and it's Requirement

Energy in any form is necessary to meet many of our requirements. It's varying definitions shows the importance of energy in our life.

1.2.1 Energy

Energy is the capacity of doing work, in fact it is a means for performing activities which are directly or indirectly related to human being.

1.2.2 Fuel

Available sources of energy are sometimes referred to as fuel. The term originally came into use to describe energy sources and to show the phenomenon of burnt oil, coal, wood, oil etc.

1.2.3 *Principal energy form*

There are six forms of principal energy:

1. Heat
2. Mechanical
3. Electrical
4. Electromagnetic wave (Light, Radio waves)
5. Chemical
6. Nuclear

1.2.4 *Energy sources*

Principally, there are two types of energy sources on the earth

1. Conventional Energy Sources
2. Non Conventional Energy Sources

There are five ultimate sources of useful energy for mankind

1. The sun.
2. The motion & gravitational potential of sun, moon & earth.
3. Geothermal energy from cooling, chemical reaction & radioactive decay on the earth.
4. Nuclear reaction on the earth.
5. Chemical reaction from mineral sources.

The conventional energy sources are derived from sources (1) Fossil fuels (2) Hot rocks, (3) Nuclear and (4) Chemical reaction.

Where as non conventional energy sources is derived from sources (1) Sun (2) Motion & gravitational forces and (3) Geothermal energy or molten megna.

1.2.5 *Definitions*

(a) *Conventional Energy:* It is energy obtained from static storage of energy sources (fossil fuel, hot rocks, nuclear and chemical) which remain static bound in position until or unless it is released by human interaction.

Examples are coal, oil & natural gas. Hence initially the energy remains in static nature which otherwise reacted by human activities and converted in kinetic form and there by energy is used. These sources in present context are known as finite sources of energy or non renewable. Example are petroleum product, coal etc.

(b) *Non Conventional Energy:* It is the energy obtained from the natural sources, which continuously or repetitive from a current in the environment, which can be easily be trapped in useful form.

Examples are solar energy, wind energy, geothermal energy and bio energy. These sources also known as renewable energy. Comparison of these sources is given in Table 1.1 and the definition is illustrated in Fig 1.1

Table 1.1. Comparison of Conventional & Non-Conventional Energy Sources

S. No.	Parameters	Conventional Energy Supplies	Non-Conventional Energy Supplies
1	2	3	4
1.	Source	Static stores in earth	Natural Sources
2.	Examples	Coal, Oil, Gas	Wind, Solar, Biomass, Tidal
3.	Nature of Occupancy	Static store of energy	A current of energy
4.	Normal state	Remain in bound position	Continuously or Repeatatively released
5.	Initial Intensity	Released @ 100 kw/sqm & more	Low intensity, dispersed 300 W/sqm or less
6.	Life Time of Supply	Finite	Infinite
7.	Cost at Sources	Increasingly expensive	Free
8.	Conversion process/ Utilization device/ Transmission & Distribution media	Established / Commercialized	Under R&D stages, certain items are established

Contd...

Table 1.1. Contd...

9.	Cost of equipment	Moderate	Quite high in present context
10.	Variation & Control	Steady, best controlled by adjusting source with feed back control	Fluctuating, best controlled by change of load using feed forward control
11.	Location for use	General use	Site & location specific
12.	Scale of production	Large scale	Small scale
13.	Skills for production	Strongly available	Under R&D stage
14.	Skills for utilization	Strongly available	Under R & D stage
15.	Context of utilization	Urban, centralized	Rural, decentralized
16.	Dependence	System dependent on outside inputs	Self-sufficient encouraged
17.	Safety	Most dangerous when faulty	Usually safe when out of action
18.	Pollution & Environmental damages	Environmental pollution common	Usually little harmful to the environmental especially at moderate scale
19.	Ecology damage	Permanent damage common from mining	Hazards from excessive wood burning/ soil erosion from excessive biofuel use.
20.	Esthetics	Small structure may produce little aesthetic difficulty	Local perturbations may be serious.
21.	Economics	Availability is costly, but harnessing is within economic limit	Availability free, but harnessing is quite costly.
22.	Over benefit in present context	Finite source, may not remain in future	Infinite, locally available. No danger on source availability

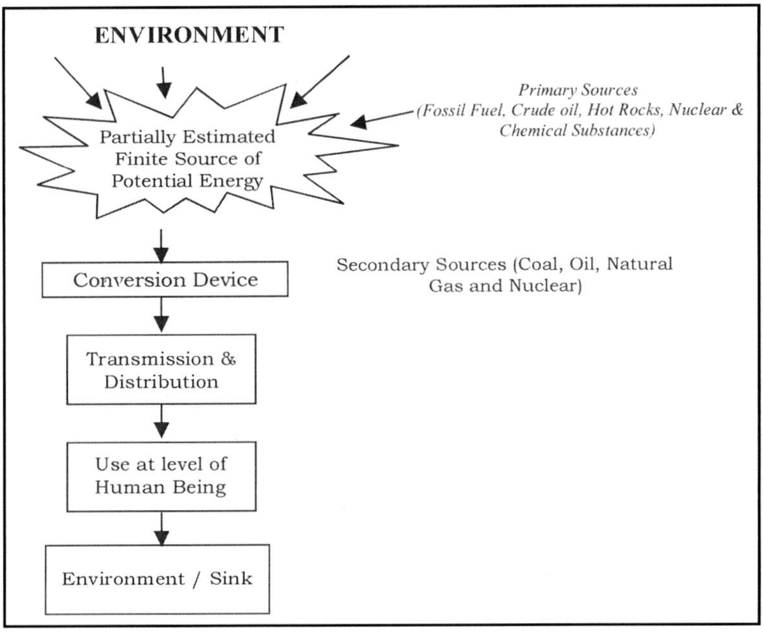

(a) Conventional Energy Sources

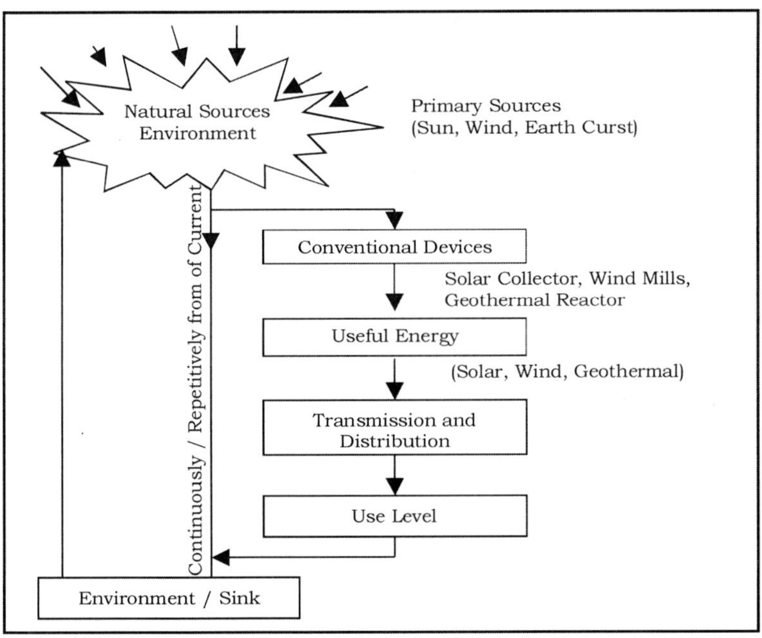

(b) Non Conventional Energy Sources

Fig. 1.1. Conventional & Non Conventional Energy Flow Pattern

1.3 Reforms in the Energy Sector

Reforms in the energy sector were initiated to supplement the Government's efforts in the development of the sector and to make it more efficient. The Government has been endeavouring to provide a policy environment that encourages free and fair competition in each element of the energy value chain and attracts capital from all sources-public and private, domestic and foreign. Encouraging such capital formation is crucial for India's to meet its energy needs. Significant progress has been made in establishing independent and transport regulatory authorities in the power sector to facilitate the rationalization of electricity tariff as well as to encourage competition while protecting the interests of all stakeholders.

The thrust of the reforms has been to deregulate the prices of commercial energy resources (which, until recently, were entirely administered), increase competition through institutional, legislative and regulatory reforms and reduce subsidies.

1.3.1 Energy priorities for developing countries

Achieving development goals in rural areas where poverty conditions are currently the greatest will require greater access and improved energy services as a means to reach education, health, water and other goals. In particular, expanding energy services is a means to generate increased employment and income generating opportunities – and is therefore a pre-requisite to increased value adding activities in rural areas. For the policy-maker, helping create a sustainable energy pathway will require broad societal consensus around the strategic choices of economic, environmental and social development. Depending on their current state of development and priorities, countries will likely pursue different paths towards a variety of sustainable development options.

1.3.2 Access and affordability

Access to commercial energy will mean that energy must be available at prices which are both affordable (low enough for access by the poorest people) and sustainable (prices which cover the real costs of energy production, transmission and distribution to support the financial ability of companies to maintain and develop energy services and which internalize the social/environmental costs associated with these activities, if practicable). The best way to ensure that a growing number of people will be able to afford commercial energy is to accelerate economic growth and assist those with the lowest income to become more wealthy.

1.3.3 Energy and the environment

Environment is one of many factors that societal decision-makers account for when considering energy choices. For example, energy production and consumption can raise environmental challenges, including land use, global climate change, water and urban air quality.

1.3.4 Energy efficiency

Energy efficiency is essentially using less energy to provide the same service. In this sense, energy efficiency can also be thought of as a supply resource – often considered an important, cost effective near- to mid-term supply option. Investments in energy efficiency can provide *additional* economic value by preserving the resource base and (especially combined with pollution prevention technologies) mitigating environmental problems.

Improvements in energy efficiency can produce direct environmental benefits in a number of ways, not only reducing pollution but also delaying the need to develop new fuel resources. In addition, energy efficiency improvements can considerably reduce the cost of pollution abatement. Industrial, commercial, and consumer equipment today can be as much as 80 per cent more efficient than equipment installed just twenty years ago. For example, improved efficiency in any power plant can produce significant reductions in CO_2 emissions. Typically, a 1 per cent point gain in efficiency reduces CO_2 output by 2 per cent.

Efficiency can also be increased at the point of consumption. For example, compact fluorescent lamps are available which use 75 per cent less electricity than conventional light bulbs, and last much longer, while providing light of a similar quality. Although cost-effective over the lifetime of the bulb, compact lamps are not used everywhere due to the greater initial cost.

This is an example of one of the major barriers to greater energy efficiency - capital outlay. The ability to use the full range of market based energy and energy technology resources, along with cleaner technologies and fuel systems, will help drive the innovation needed to optimize business and societal activities within the framework of sustainable development.

1.4 Role of Energy in Economic Development

Humanity has searched a variety of energy sources in order to get greater comfort, enhanced security from want and the satisfaction of wants through increasingly elaborate artifacts. History reveals three energy areas: wood, coal and petroleum fuel are being predominately used.

The use of wood as a fuel for space heating was even used during prehistoric times, but it employment as a fuel for manufacturing or industrial purpose probably dates back less than 50,000 years. Coal began to be employed on a large scale in Europe only after the elimination of forest cover from areas of intensive fuel use. In fact industrialized countries used wood as major energy sources only upto the eighteenth century. During the nineteenth century, coal replaced wood, wood and coal, coupled with the development of steam generation technology, enabled a massive substitution of human labour by inanimate energy.

When crude oil was first marketed on large scale in the 1860's, it did not immediately begin to crowd coal out of the market. However, the lubricants drived from petroleum were unique suitable to the needs of high speed machinery and many advances in mechanical engineering would have been impossible without the improved lubricants derived from crude oil. Eventually it began to complete more effectively with coal when its value as a boiler fuel was recognized and its costs became competitive with those of coal. However, it was not until the development of internal combustion engine that crude displaced coal, in the first part of the twentieth century. Natural gas began to complete with crude as a source of nation wide energy supplies in the 1930's when the technology of long distance low cost gas pipeline was developed. For a number of years gas consumption grew more rapidly that that of coal and gas displaced oil in same market because of its superior convenience, cleanliness and economy.

In fact, wood, coal and petroleum energy played a vital role in economic development. As an economy develops, it demands locally available more energy quantity wise and quality wise in more convenient forms. In general, the more rapid the rate of economic growth, the more rapid the replacement of solid fuels by fluid fuels.

In earlier years petroleum product made considerable contributions to the economic growth in direct as well as indirect ways. Later on, with the invention of electric power and the internal combustion engine, the entire population got enhanced mobility. In the farm, the IC engine made it possible to turn out more foodstuffs. Further, electrification in turn spurred the development of new industrial process which greatly enhanced labour productivity. Similarly such advances are more evident in the home, with the advent of the electric refrigerator, microwave oven, stove, dishwasher, radio and many other home appliances.

1.5 Indian Energy at a Glance

India ranks sixth in the world in terms of energy demand accounting for 3.5 per cent of world commercial energy demand. With a gross domestic product (GDP) growth of 8 per cent set for the Tenth Five-Year Plan (2002-07), the energy demand is expected to grow at 5.2 per cent. Although, the commercial energy consumption has grown rapidly over the last two decades, a large part of India's population does not have access to it. At 479 kg of oil equivalent (kgoe), the per capita energy consumption is also low even compared to some of the developing countries.

India is fortunate to be endowed with both exhaustible (particularly coal) and renewable energy resources. Despite the resource potential and the significant rate of growth in energy supply over the last few decades. India faces serious energy shortages. This has led to reliance on increasing imports for meeting the demand of oil and coal. As per current projections, India's dependence on oil imports is expected to increase. The demand of natural gas also outpaces supply and efforts are being made to import natural gas in the form of liquefied natural gas (LNG) and piped gas. The power sector has also been experiencing severe shortages.

The development strategies proposed for the sector includes increasing the production of coal and electricity, accelerated exploration for hydrocarbons, equity oil abroad, introduction of reforms through restructuring/deregulation of the energy sector to increase efficiency, demand management through introduction of energy efficient technologies/processes and appliances. The process of producing, transporting and consuming energy has a significant impact on the environment. Pollution abatement processes would form an important part of the development of energy sector.

1.5.1 Energy scenario

Primary commercial energy demand grew almost three-fold at an annual rate of 6 per cent between 1981 and 2005, to reach 314.7 million tones of oil equivalent (MTOE). India's incremental energy demand for the next decade is projected to be among the highest in the world, spurred by sustained economic growth, rise in income levels and increased availability of goods and services.

India's commercial energy demand is expected to grow even more rapidly than in the past as it goes down the reform path in order to raise standards of living. A large part of India's population does not have access to commercial energy. The 479 kgoe per capita total energy consumption is only about 20 per cent of the global average in 1997 and compared poorly with the per capita consumption of Thailand (1,319 kgoe), Brazil (1,051 kgoe) and China (907 kgoe).

1.5.2 Non-commercial energy resources

More than 60 per cent of Indian households depend on traditional sources of energy like fuel wood, dung and crop residues for meeting their cooking and heating needs. Out of the total rural energy consumption, about 65 per cent is met from fuel wood. Fuel wood consumption during 2001-02 is estimated at 223 million tones, 180 million tones of which is for household consumption and the balance for cottage industry, big hotels etc. The consumption of animal dung and agro-waste is estimated at 130 million tones, which does not include the wet dung used for biogas plants. It is assumed that the wet dung used as manure is being diverted to biogas plants as these plants, in addition to providing a cleaner fuel, also supply enriched manure.

1.6 Trends of Economic Growth and Energy Use

The projected requirement of commercial energy is estimated at about 412 MTOE and 554 MTOE respectively in 2007 and 2012, the commercial energy demand is estimated to grow at an average rate of 6.6 per cent and 6.1 per cent respectively during the period 2002-07 and 2007-12. However, the demand may be less by 5 per cent and 10 per cent during 2006-07 and 2011-12 respectively due to increasing use of information technology (IT) and prevalence of e-Commerce, which will mainly affect the demand of energy in transport sector.

Estimated Energy Demand in India

Primary	Unit	Demand (In Original Units)		Demand (MTOE)	
		2006-07	2011-12	2006-07	2011-12
Coal	Mt	460.50	620.00	190.00	254.93
Lignite	Mt	57.79	81.54	15.51	22.05
Oil	Mt	134.50	172.47	144.58	185.40
Natural Gas	BCM	47.45	64.00	42.70	57.60
Hydro Power	BKwh	148.08	215.66	12.73	18.54
Nuclear Power	BKwh	23.15	54.74	6.04	14.16
Wind Power	BKwh	4.00	11.62	0.35	1.00
Total Commercial Energy				411.91	553.68
Non-Commercial Energy				151.30	170.25
Total Energy Demand				563.21	723.93

mt: Million Tonnes; BCM: Billion Cubic Meter; Bkwh: Billion kilo watt hour

1.7 Energy Resources

India's energy use is mostly based on fossil fuels. Although the country has significant coal and hydro resource potential, it is relatively poor in oil and gas resources. As a result it has to depend on imports to meet its energy supplies. The geographical distribution of available primary commercial energy sources in the country is quite skewed, with 77 per cent of the hydro potential located in the northern and north-eastern region of the country. Similarly, about 70 per cent of the total coal reserves are located in the eastern region while most of the hydrocarbon reserves lie in the west.

Regional Distribution of Primary Commercial Energy Resources

Region	Coal (bt)	Lignite (bt)	Crude Oil (mt)	Natural Gas (BCM)	Hydro Power (TWH)
Northern	1.06	2.51	0.03	0.0	225.00
Western	56.90	1.87	519.47	516.42	31.40
Southern	15.46	30.38	45.84	80.94	61.80
Eastern	146.67	0	2.19	0.29	42.50
North-Eastern	0.89	0	166.17	152.00	239.30
Total	220.98	34.76	733.70	749.65	600.00

bt: billion Tonnes; BCM: Billion Cubic Meter; TWH: Trillion Watt Hours; mt: Million Tonnes

1.7.1 Coal

The geological coal reserves of the country are estimated at 220.98 billion tones (bt) as on January 2001. Out of this, proven reserves are 84.41 bt, while 98.55 bt are indicated reserves and 38.02 bt are inferred reserves. Coal continues to remain the principal source of commercial energy accounting for nearly 50 per cent of the total supplies. About 70 per cent of the power generated is coal and lignite based and this trend is likely to continue in the coming future.

India has an estimated 1000 billion cubic meters of Coal Bed Methane (CBM), which is likely to emerge as a new source of commercial energy in the country. A demonstration project is under implementation with financial support from the Global Environment Facility (GEF) and the United Nations Development Programme (UNDP). In April 2001, the Government announced a programme for exploration and production of CBM. Under the first round of bidding, five CBM blocks have been awarded to private companies. Apart from this, exploration work in two blocks has been awarded to two public sector undertakings (PSUs) on nomination basis. The successful implementation of these projects will facilitate exploitation of this clean source of energy.

1.7.2 Lignite

The current estimates of geological lignite reserves in India are 34.76 bt spread over Tamil Nadu and Pondicherry (87.5 per cent),

Rajasthan (6.9 per cent), Gujarat (4.9 per cent), Kerala (0.31 per cent) and Jammu and Kashmir (0.37 per cent). The lignite deposits in the southern and western regions have emerged as an important source of fuel supply for states like Tamil Nadu, Rajasthan and Gujarat. Over the years, considerable emphasis has been placed on the development of lignite for power generation. Lignite production is likely to increase from 24.3 million tones in 2001-02 to 55.96 million tonnes in 2006-07.

1.7.3 Oil and natural gas

The latest estimates indicate that India has around 0.4 per cent of the world's proven reserves of crude oil. As against this, the domestic crude consumption is estimated at 2.8 per cent of the world's consumption. The balance of recoverable reserves as estimated in the beginning of 2001 is placed at 733.70 million tonnes (mt) of crude and 749.65 billion cubic meters (BCM) of natural gas. The share of hydrocarbons in the primary commercial energy consumption of the country has been increasing over the years and is presently estimated at 44.9 per cent (36.0 per cent for oil and 8.9 per cent for natural gas). The demand for oil is likely to increase further during the next two decades. The transportation sector will be the main driver for the projected increase in oil demand. Consequently import dependence for oil, which is presently about 70 per cent, is likely to increase further during the Tenth and Eleventh Plans.

India has about 0.4 per cent of world's natural gas reserves. Initially the gas reserves had been developed largely for use as petrochemical feedstock and in the production of fertilizers, but gas is increasingly being used for power generation, industrial applications and more recently in the transport sector. Presently the share of power generation capacity based on gas is about 10 per cent of the total installed capacity. The India Hydrocarbon Vision 2025 of the Government identifies natural gas as the preferred fuel for the future and several options are being explored to increase its supply capacity including building facilities to handle imports of liquefied natural gas (LNG) and setting up of pipelines from major gas producing deposits of gas hydrates. However the true extent of this resource and its potential for commercial exploitation is still being evaluated.

1.7.4 Hydro electric potential

The key advantage of hydroelectric power is the ability to store energy and the flexibility of its use during peak load periods. India is endowed with economically viable hydro potential. The Central Electricity Authority (CEA) has assessed India's hydro potential to be about 148,700 MW of installed capacity. The hydroelectric capacity currently under operation is about 26,000 MW and 16,083 MW is under various stages of development. The CEA has also identified 56 sites for pumped storage schemes with an estimated aggregate installed capacity of 94,000 MW. In addition, a potential of 15,000 MW in terms of installed capacity is estimated from small, mini and micro hydel schemes.

1.7.5 Nuclear resources

Nuclear energy has the potential to meet the future needs of electricity demand in the country. The country has developed the capability to build and operate nuclear power plants observing international standards of safety. The current installed capacity of nuclear power plants is 2,860 MW accounting for 2.8 per cent of the total installed capacity of the country. The Nuclear Power Corporation of India Ltd. (NPCIL) proposes to increase the installed capacity to 9,935 MW by 2011-12. The future strategies focus on a three-stage nuclear power programme for the optimal utilization of the available nuclear energy resources. The first stage of 10,000 MW is based on pressurized heavy water reactor (PHWR) using indigenous natural uranium resources. The second stage is proposed to be based on fast breeder reactor (FBR) technology using plutonium extracted by reprocessing of the spent fuel the first stage. In the third stage, the country's vast thorium resources will be utilized for power generation.

1.7.6 Renewable sources of energy

India is endowed with abundant natural and renewable resources of energy viz., sun, wind and biomass. The country has been able to achieve significant capacity addition of 1,367 MW through wind farms and ranks fifth in the world after Germany, United States, Spain and Denmark in the generation of wind energy. The available renewable resources need to be exploited by giving a commercial orientation, wherever possible. It may be necessary to continue with subsidies in the case of socially oriented programmes to meet the energy requirements of rural areas, particularly, remote villages, which may be difficult to service through the conventional power grid in the near future.

Renewable Energy Sources Potential

Source/Technology	Units	Potential/ Availability	Potential Exploited
Biogas Plants	Million	12	3.22
Biomass-Based Power	MW	19,500	384
Efficient Wood Stoves	Million	120	33.86
Solar Energy	MW/Sq.Km	20	1.74
Small Hydro	MW	15,000	1,398
Wind Energy	MW	45,000	1,367
Energy Recovery from Wastes	MW	1,700	16.2

1.8 Energy Production

The country has seen an expansion in total energy use during the last five decades, with a shift from non-commercial to commercial sources of energy.

Trends in Commercial Energy Production

	Units	Production					
		1960-61	1970-71	1980-91	1990-91	2001-02*	2006-07**
Coal	Mt	55.67	72.95	114.01	211.73	325.65	405.00
Lignite	Mt	0.05	3.39	4.80	14.07	24.30	55.96
Crude Oil	Mt	0.45	6.82	10.51	33.02	32.03	33.97
Natural Gas	BCM	-	1.44	2.35	1.79	29.69	37.62
Hydro Power	Bkwh	7.84	25.25	46.54	71.66	82.80	103.49
Nuclear Power	Bkwh	-	2.42	3.00	6.14	16.92	19.30
Wind Power	Bkwh	-	-	-	0.03	1.70	4.00

*Anticipated, **Projections for the terminal year of the Tenth Plan

Coal production is likely to grow at an annual rate of 4.46 per cent during 2002-07 (compared to 2.4 per cent annual growth rate during 1997-02) to touch 405 mt in the terminal year, 2007. As against this, the coal demand in that year is estimated at 460.50 mt. Part of the gap is proposed to be met through import of both coking and non-coking coal. About 70 per cent of the projected demand is for public sector utilities. A substantial expansion in the domestic coal production is, therefore, needed to meet the requirements of the targeted generating capacity additions envisaged during the Tenth and Eleventh Plans.

The current domestic production of crude oil caters to nearly 30 per cent of the demand and is likely to marginally increase from 32.03 mt in 2001-02 to 33.97 mt in 2006-07. As against this, the demand for petroleum products, projected as 99.13 mt in 2001-02, is estimated to grow at the rate of 5.7 per cent a year to touch 134.6 mt in the terminal year of the Tenth Plan and 172.5 mt in the terminal year of Eleventh Plan.

India's natural gas production reached a level of 29.69 BCM in 2001-02. The projected domestic production of natural gas in 2007 is 37.62 BCM. The country has been able to meet the demand with the available domestic production till recently. However, the demand is likely to grow rapidly in the near future. A number of projects for setting up of LNG terminals have been approved by the Government to bridge the demand-supply gap. Four LNG terminals at Dabhol, Dahej, Hazira and Cochin are in advanced stages of development and are likely to be completed by the end of the Tenth Plan.

Significant hydro and nuclear generation capacity is likely to be added during 2002-07 period. The capacity addition programme includes 16,083 MW from hydel power plants and 1,300 MW from nuclear power plants. In addition, 2,000 MW of energy is planned to be harnessed from wind farms.

1.9 Comparative Trends

Though coal production increased about three times from 114 mt in 1980-81 to 325 mt in 2001-02, the share of coal in total energy supplies has declined from a level of 58.9 per cent to 51.1 per cent. This could be partly due to the increase in the share of inferior grade coal in over-all coal production. The primary reason, however, is that the share of hydrocarbons in the total energy consumption of the country has been increasing over the years and is currently estimated at 44.9 per cent as compared to 37.2 per cent in 1980-81. Net energy related imports of 87.85 MTOE in 2001-02 include the import of 75.43 mt of crude and petroleum products, 19.60 mt of coal and 1.4 Bkwh of electricity from Bhutan. The share of non-commercial sources in the total primary energy supply is 31.8 per cent in 2001-02, down from 53.1 per cent in 1980-81.

Trends in Supply of Primary Commercial Energy

	Supply					
	1953-54	1960-61	1970-71	1980-81	1990-91	2001-02*
Coal	23.62	35.64	36.48	56.96	94.68	133.89
Lignite	-	0.01	0.81	1.23	3.34	6.52
Crude Oil	0.19	0.46	7.01	10.79	33.92	32.03
Natural Gas	-	-	0.60	1.41	11.73	26.72
Hydro Power	0.24	0.67	2.17	4.00	6.16	6.37
Nuclear Power	-	-	0.63	0.78	1.60	5.15
Wind Power	-	-	-	-	-	0.14
Total	24.05	36.78	47.67	75.19	151.43	210.82
Net Imports	2.20	6.04	12.66	24.63	31.69	87.85
Commercial Energy Supply	26.25	42.82	60.63	99.82	183.12	298.67
Primary Non-Commercial Energy Supply	64.13	74.38	86.72	108.48	122.07	139.02
Total Primary Energy Supply	90.38	117.20	147.05	208.30	305.19	437.69

1.10 Energy Imports

India is emerging as a large importer of crude and is planning to import LNG during 1977-02. If the present trend continues, India's oil import dependency is likely to grow beyond the current level of 70 per cent. Future strategies world focus on increasing exploration activities to enhance the level of recoverable reserves of the country.

Coal imports account for only about 5.6 per cent of the total domestic consumption in the current year. The steel sector has been importing coking coal mainly for blending with domestic coal to obtain the desired quality for steel production. The cement industry and coastal power stations are importing non-coking coal.

Share of Net Energy Imports in Primary Commercial Energy Supply

per cent				
Year	Coal	POL	Electricity	Total
1980-81	0.25	25.45	-	25.70
1990-91	2.22	15.56	0.07	17.85
2001-02	4.12	25.25	0.04	29.41

2

Integrated Energy Planning

2.1 Introduction

Energy is one of the main input for rural development and economic development. The greater part of the world population lives at low standard and a quarter of the population or about 1.5 billion, have an extremely low standard of life. Energy is one of the essential requirement of the quality of life the people. The industrial development or even large scale food production, water supply, textile manufactures & construction of building and other recreation items are not possible without adequate supply of energy. Therefore, one of the most important tasks is to provide adequate, reliable, convenient and efficient supply of energy, if quality of life of the people is to be raised from its present level of grinding poverty. Table 2.1 project the global primary energy supply in six scenarios i.e A1 - its strong emphasis on oil and natural gas use. A2 - which is coal-intensive. A3 - which emphasises the roles of natural gas, new renewables and nuclear in averting serious problems from emissions. B - Middle course. C1- with its emphasis on energy efficiency improvements, new renewables (especially solar in the longer run), but with nuclear power phased out by 2100 because unable to satisfy its critics and C2 - where nuclear power plays an expanding role.

Table 2.1: Projections of the Global Primary Energy Supply

Energy Sources		(Gtoe) Giga tonnes (10^9) of oil equivalent					
	1990	2050					
		A1	A2	A3	B	C1	C2
Coal	2.2	3.8	7.8	2.2	4.1	1.5	1.5
Oil	3.1	7.9	4.8	4.3	4.0	2.7	2.6
Gas	1.7	4.7	5.5	7.9	4.5	3.9	3.3
Nuclear	0.5	2.9	1.1	2.8	2.7	0.5	1.8
Hydro	0.4	1.0	1.1	1.1	0.9	1.0	1.0
New Renewables	0.2	3.7	3.8	5.7	2.8	3.8	3.2
Traditional Biomass	0.9	0.8	0.7	0.8	0.8	0.8	0.8
Total	**9.0**	**24.8**	**24.8**	**24.8**	**19.8**	**14.2**	**14.2**

Broadly speaking of the total energy used in rural India, the share of human and animal power is 15 percent, non commercial sources (firewood, agricultural waste, cattle dungs etc.) 65 percent and commercial energy 20 percent. Share of commercial energy constitute about 23 percent in agriculture (fertilizer 14 percent, electricity 6 percent and diesel 3 percent), nearly 2 percent in domestic cooking and 100 percent domestic lighting (kerosene 87 percent and electricity 13 percent). Most of significantly, transportation and rural industries sector solely depends on animate and non commercial sources of energy. The Fig. 2.1 and 2.2 shows the global total energy consumed and total fossil fuel consumed respectively.

2.2 Rural Energy Scenario

In the present rural energy scenario, one of the important question is what are the needs and which energy resources might be employed to satisfy rural requirement. These basic energy requirements in rural area are summarized in Table 2.2. In terms of energy forms, there energy requirement can be classified as:

(1) Heating or Cooling

(2) Mechanical Shaft Power

(3) Lighting

(4) Sound (for communication, radio etc.)

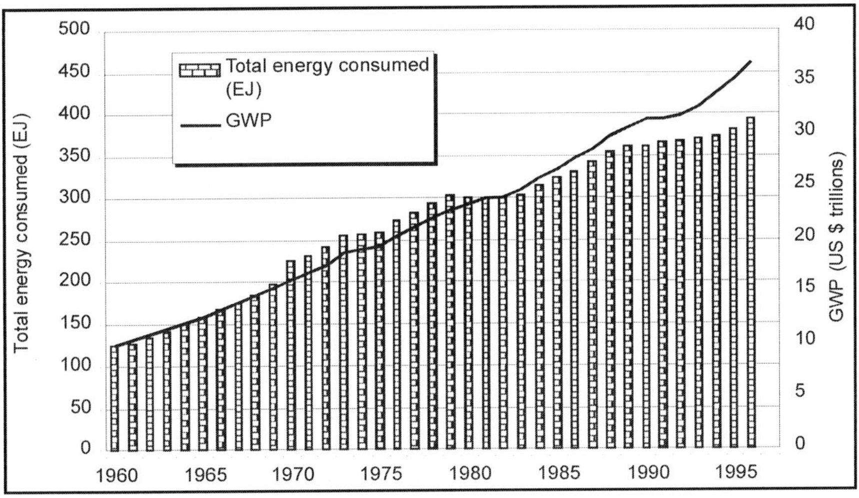

Fig. 2.1 Global Energy Consumed

EJ (exajoules) = 10^{18} joules.

GWP = gross world product

Source: Energy Section, Statistical Office, United Nations, New York, New York, U.S.A. International Energy Agency. 1998. Energy Statistics & Balances of Non-OECD Countries 1998 Edition. Paris.

Brown, L.; M. Renner; and B. Halweil. 2000. Vital Signs 2000. W.W Norton & Company: New York

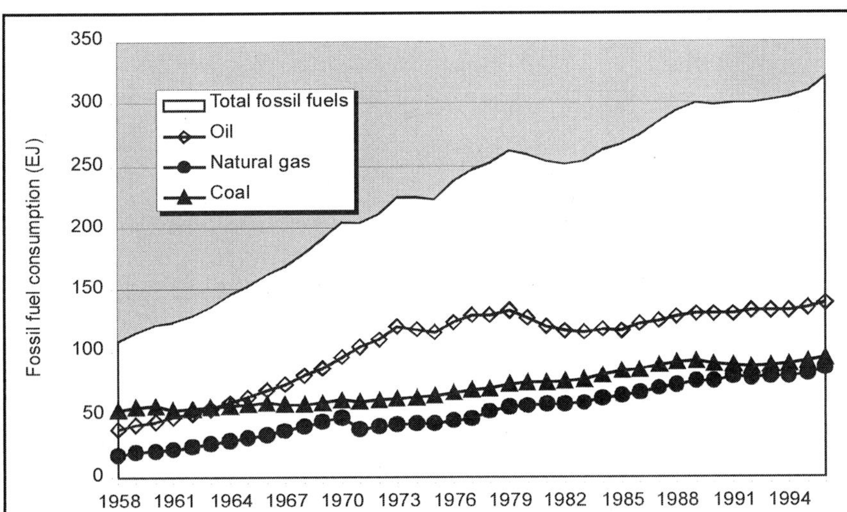

Fig. 2.2 Global Fossil Fuel Consumed

Source: Energy Section, Statistical Office, United Nations, New York, New York, U.S.A. International Energy Agency. 1998. Energy Statistics & Balances of Non-OECD Countries 1998 Edition. Paris

Table 2.2 Energy Requirements in Rural Areas

S. No.	Activities	Common Sources Used	Common Devices Used
1.	Domestic		
	Heating/Cooling	Firewood, Agri-waste, dung et.	Wood/Charcoal, Stove
	Cooking	-do-	-do-
	Lighting	Kerosene , Electricity	Traditional lamp, Incandescent & Fluorescent lamp
2.	Agricultural Operations		
	Land Preparation	Human/Animal Power Diesel	Different Agricultural implements and tractors etc.
	Sowing, Weeding and Harvesting	-do-	-do-
3.	Crop Processing & Storage		
	Drying	Solar Energy	Open/Improved sun drying
	Grinding	Human/Animal Power	Flour mills
	Refrigeration	Electricity, Diesel, Petrol	Refrigerator
4.	Small Industry		
	Power Machinery and Process Heat	Electricity, Diesel, Petrol	Different Machineries
5.	Water Pumping		
	Domestic Use	Human Power	Conventional Methods
	Cattle	Animal Power	Tube Well
	Irrigation	Electricity, Diesel & Petrol	Ground Pump set
6.	Transport	Animal Power, Diesel, Petrol	Cart, Automobile

Contd...

Table 2.2. Contd...

7.	School	Human Power & Electricity	Traditional
8.	Communication	Human Power & Electricity	Telephone, Mobile etc.

The fundamental problem of rural energy planning is energy resources, requirements & the optimum energy flow to meet the energy needs with the available resources. Hence, in the planning it is essential to consider the feasible technological options, including the traditional ones, for satisfying these energy needs with available sources (conventional & non conventional) excluding exploitation of improved method to increase the resources.

2.3 Rural Energy Planning

In most of the rural areas, different energy sources and devices are available, which are capable to meet their energy requirement. However, these sources contribute to well being people only if these sources are properly harnessed & utilized. To attain specific goals a proper balance is needed among, technical, social & economic benefits derived from these sources and the social & economic costs required for operation & maintenance. To achieve this balance, planning is needed.

To implement a techno economically sound rural energy system, generally it is necessary to follow planning process, where by a choice of the best course of action is made from a set of alternatives or when the rural energy system already chosen. It is necessary to select the best approach for the implementation of the project from a range of possible methodologies or devices. Regardless of how it is structured, this is planning process.

In addition to this, if energy flow from its origin to end use is to be evaluated, than one can see the different intermediate steps, which are reliable and convenient. But these steps are not efficient & a tremendous amount of energy is being lost before coming to end use activities. The Fig 2.3 illustrated the energy flow in present context. From the figure, it is clear that there is greater scope of improvement in making energy sources efficient & for this proper planning is required.

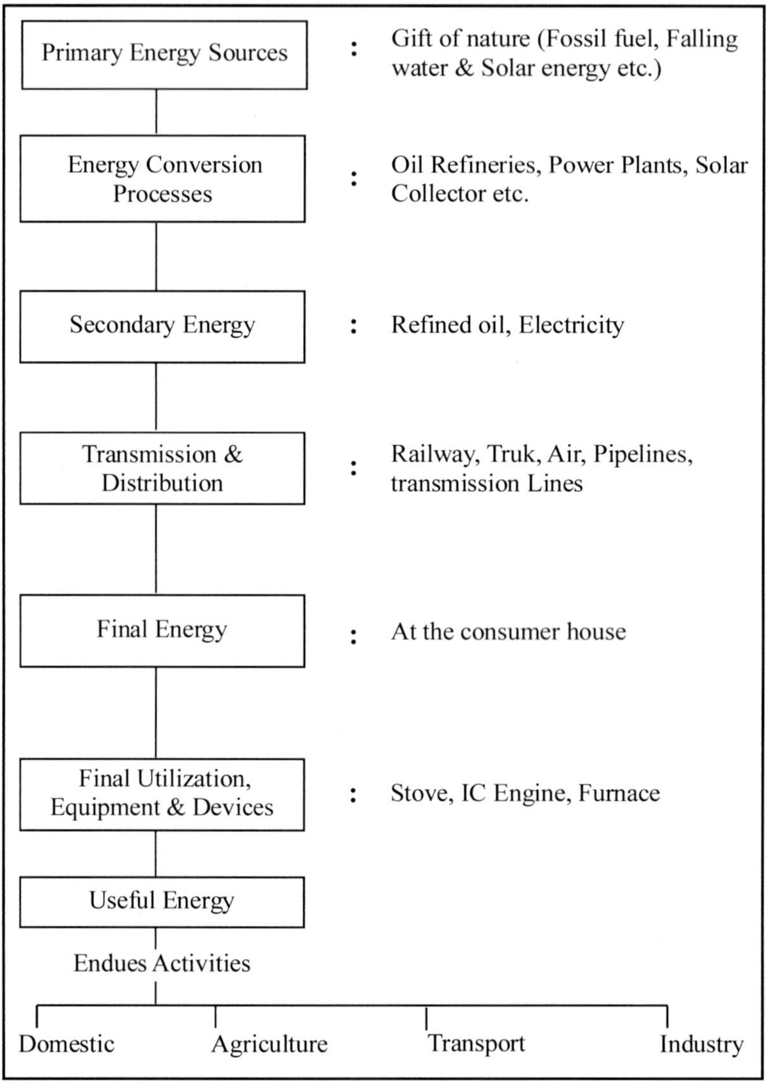

Fig. 2.3 Energy Flow Pattern in Present Context

Planning beings with a dialogue where by local people of rural areas assess their needs, define their actual demands in terms of goal & objectives, and agree on methods of meeting out the demands. The results of this dialogue are a consensus which has emerged from discussion among community members and endorsed by implementing authorities and decision makers. This shared responsibility & understanding of an approach to an objective or problem. The rural energy planning is especially critical for two main reason:

1. In rural energy planning, technical, economic & social issues are so closely interlinked.
2. In rural energy planning, long term period of sustainability is often needed for energy sources to yield optimum and continuous supply.

In many instances a planning process can be time consuming. However, without effective communication among planners, users, economists, politician, implementers and other involved in the process a rural energy management practices is likely to be delayed in implementation or neglected, once the plan has been implemented. A commitment to share in the planning process does not necessarily guarantee that a plan will succeed, but it is a requirement for the support needs to initiate & maintain rural energy related development projects. Hence community participation is must for getting success in rural energy planning process. Rural energy system will flourish only if the people participate in it.

2.4 Energy Use in Rural Areas

At least half of gross energy consumption in most developing countries occurs in rural areas. The bulk of this energy is derived from locally available traditional energy resources. It provides energy services via rudimentary technologies, which are limited in capability and generally inefficient. Modern, commercial energy sources, such as electricity and petroleum-based fuels, generally provide a small part of the energy consumed by rural people, mainly because of supply and affordability constraints. In rural areas, in addition to households' need for a minimum amount of energy, there is an increasing demand for energy in the provision of rural services such as water supply, health care and education, and for productive activities such as agriculture and small industries. Ideally, all these needs should be met in an efficient, cost-effective and environmentally sustainable manner. In many cases, this is really only possible through a transition from the use of inefficient traditional energy sources and technologies, to more efficient technologies and sources of energy. In so much as the facts are available, this chapter describes patterns of energy use in the rural areas of developing countries.

2.4.1 General characteristics of rural energy use

Energy use in rural areas can be broken down into the household, agricultural and small-scale rural industry sub-sectors and services.

Since the amount of energy use for services (health clinics, schools, street lighting, commerce, transport, etc.) is generally quite small in rural areas, it is often included in the rural industries sector.

Agricultural activities consume from 2 to 8 per cent of the total, depending on levels of mechanisation, mainly in the form of commercial energy used to power mechanical equipment and irrigation pump-sets. In general these statistics do not include human and animal power that provide the bulk of agricultural energy input for the basic agricultural activities.

Commercial energy, mainly kerosene and electricity where available, is mainly used for lighting, which on average constitutes about 2 to 10 per cent of total rural consumption. Small amounts of electricity are used to operate radios, television sets and small appliances in electrified villages. This has serious implications for many rural electrification projects. Electricity demand curves in many rural areas are characterised by high peaks in the early evening hours and a low overall consumption, which means high investment in peak capacity installations and low returns.

The energy consumption of rural industries, including both cottage industries and village level enterprises, amounts to less than 10 per cent of the rural aggregate in most countries. Woodfuel and agricultural residues constitute the principal sources of supply for these activities, with electricity sometimes providing some motive power. Religious festivals, celebrations, burials and other occasional functions may also consume large amounts of fuel.

2.4.2 Rural household energy use

An enormous variety of household energy consumption patterns are to be found in the rural areas of developing countries. Remedies for rural energy problems have to be based on a good understanding of dramatically varying local conditions and the key variables that affect the levels of demand and the types of energy sources used. However, energy planners seldom if ever have access to information at this level of detail and this in turn has profound implications for the planning of rural energy development.

Households use energy for many purposes. How much they consume and the types of energy source they use depend on a variety of factors, which include the availability of energy sources and the personal or cash costs entailed in obtaining and using them. However, they also include many factors that can only be understood by looking at the real needs and behaviour of energy consumers. Analysis of household energy must, therefore, be rooted in a sensitive approach not only to issues of supply, but more importantly, to those of energy demand. A household's total energy consumption and mix of energy sources is the result of its attempt to meet its various energy needs by using its labour, cash and energy using appliances in the most efficient way possible. The micro-perspective of each end-user is the driving force behind the sector's use of energy and is the key to the opportunities for change in demand and supply patterns.

Among the poorest families in most developing countries, cooking (and space heating depending on the climate) accounts for between 90 and 100% of energy consumption. The remainder of the energy consumed is for lighting provided either by the cooking fire, kerosene lamps, candles or electric torches. At higher income levels, better lighting is one of the first energy services sought to improve living standards and, frequently, to extend the working day. At still higher incomes, water heating, refrigeration and cooling begin to play an important role.

In addition, at these higher income levels, the need for space heating may well decline since dwellings may be better constructed.

2.4.3 *Cooking*

The amount of energy used for cooking depends on many factors such as the type of food cooked, the number of meals cooked, household size, the specific combination of energy source and cooking equipment employed (type of stove, cooking pans), and the way in which cooking devices are used. Cooking of staples and other foods varies greatly both in terms of time and the rate of heat input required. For example, rice is usually boiled or steamed for 20 to 30 minutes, while kidney beans may be boiled for four hours or more. Field measurements of the specific fuel consumption to cook various staple foods found that rice for an average household takes between 12 and 38 MJ/kg to cook, while beans can take up to 225 MJ/kg. These measurements were based on cooking with wood fires.

Since diets include food other than staples, another useful indicator is cooking energy consumption per person-meal or per person-day. Another research project found that daily cooking energy consumption per capita varied from 11.5 to 49 MJ, based on field measurements.

Despite a wide range of locations and conditions the range of consumption is quite small. In all these cases food was cooked predominantly on an open wood fire. However, the lower figures are those applying to efficient wood or charcoal stoves and modern energy sources. In households where modern cooking energy sources and equipment are used, and the preparation of partially cooked food is common, specific fuel consumption is found to be in the region of 2 to 3 MJ/capita/day.

The effect of different cooking technologies when one type of meal is cooked can be seen from the following table, based on field tests conducted in Fiji. If we use the energy consumed using the Primus stove as a point of reference it can be seen that cooking the same meal can take up to 14 times as much energy when using a chulha stove as illustrated in Table 2.3. Interestingly an open wood fire was found to be more efficient than both the woodstoves.

Table 2.3: Energy Consumption, Relative Efficiencies and Cooking Times for a Standard Meal on Various Appliances Using Different Energy Sources

Energy Source	Appliance	Energy Consumption	Relative Energy Consumption (MJ)	Cooking Time (minutes)
Kerosene	Primus	2.5	1	70
	Wick	8.2	3	63
Charcoal	Stove	13.1	5	75
Wood	Open Fire	18.0	7	70
	Chulha Stove	35.0	14	95

Cooking is an end-use in which one finds strong and often highly specific energy source preferences. The reasons for choosing particular energy sources and cooking appliances include ease of handling and lighting, flame quality and temperature, ability to secure a fire from young children, smokiness and the taste imparted to food, as well as relative costs and the availability of energy sources and appliances. These same factors may lead households to reject 'improvements' such as more efficient stoves that do not satisfy their customs and preferences.

2.4.4 *Lighting*

Most families in rural areas of developing countries rely on the cooking fire and possibly candles and some use of an electric torch to meet all their lighting needs. Kerosene is the next most widely used lighting energy source, while for very few electricity, either from the grid, a diesel or petrol powered generator, or solar photovoltaic panels is an option.

As rules of thumb, where 90 to 100 per cent of lighting needs are met by kerosene, for low to middle income groups, consumption is roughly 1.8 to 3.6 GJ per household per year, or about 0.07 to 0.28 litres per night, with lighting periods of between two and four hours per night. Kerosene is burned either in open wick lamps, typically with a naked flame from a wick protruding from a simple jar or bottle of fuel; enclosed wick lamps, in which the wick is surrounded by a glass chimney that creates an updraft past the wick and promotes a hotter, brighter flame; or pressurised lamps which normally use a coated mantle to provide an intense white light. Except for the pressurised lamps, most kerosene lamps provide very low lighting intensities, far below those required to meet reading needs adequately.

Few kerosene lamps have efficiency better than one tenth that of electric lighting. As a result, one frequently finds that the running costs of electric lighting are much less than lighting by kerosene for an equivalent light output. This point is of great importance for energy source substitution. Since electricity almost invariably replaces kerosene for lighting and not vice versa, one might expect energy consumption to fall after the switch because of the much greater efficiency of electric lighting. However, most consumers increase their lighting standards (intensities) at the same time.

Although not yet widespread, solar photovoltaic lighting should be mentioned, as it can in some instances be a viable alternative to the more traditional lighting systems. Studies which have compared the economics of kerosene, diesel-electric and solar lighting in remote rural areas have found that solar and diesel costs are fairly close and generally lower than kerosene, assuming the same quantity of lighting for each method. Although this is likely to be the case in sunny regions where no electric grid exists and diesel fuel is expensive or hard to obtain, where these limitations do *not* exist, photovoltaic lighting is unlikely to be economic, at least at present costs. Furthermore, in the absence of subsidies or a well-structured financing mechanism, the high initial cost is bound to be an insurmountable barrier for most households.

2.4.5 *Space heating*

The main impact of space heating is not only that it raises total energy needs but also that it raises them during seasons when it is often more difficult to collect, store and dry biomass fuels.

2.5 Planning Process

The rural energy planning process follows a sequence of several phases. Although the overall process might be described in different ways, the major steps are:

1. Identifying problem & objective by the local community.
2. Establishing the criterion of acceptance with those who will carry on the rural energy management practice.
3. Evaluating the alternatives, including both improved alternatives including traditional ones and implementation methods.

Various quantitative procedures can be employed in completing the phase of a planning process. Many of these techniques are quite detailed in terms of their application. Some of these planning techniques require the use of simulation technique & computer programme. Other techniques are based on a listing of steps to be considered as the planning process proceeds. The sequence of steps in a general rural planning works is illustrated in Fig 2.4

The different steps encountered in the planning process illustrate in Fig. 2.4 are described as below.

(a) Definition of the Problem

In the rural energy planning process, the people must understand and agree on the problem of energy sources & devices. Special studies and the collection of information is needed once the problem has been identified for example, if there is a problem of cooking energy sources in a village and traditional supply of fuelwood is also out of the reasonable walking distance from the village, information on the structure of the village population and history of cooking techniques will be necessary to identify the problem. Much of this information can be obtained directly from the village and supplemented by information from developmental workers in the area. Sometimes, this informational need is called the "needs assessment" or the "needs identification" stage in planning. In this stage, a sound planning process must include the obtainment of this information at an early stage.

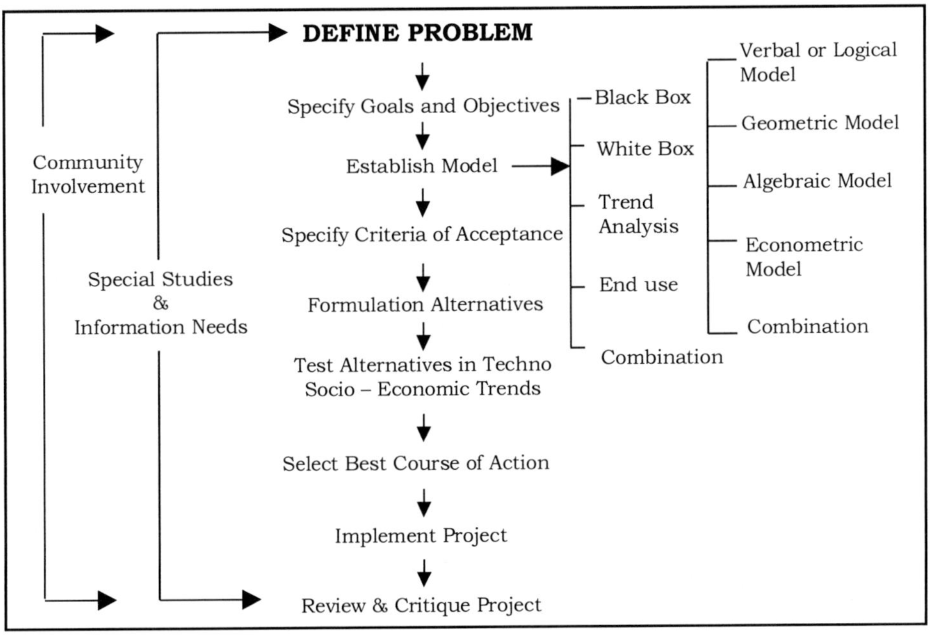

Fig. 2.4. General Planning Frame Work

(b) Specification of Goal & Objectives

The involvement of people in specifying the goal & objectives of a rural energy resources and system. In setting priorities for these goals & objectives, is critical to the planning process. When the "key" people are not involved, there is little chance that the project will be sustained over a long period of time. in the rural energy planning process, it is must to specifying goal & objectives of the problem in the present context. Say for example, if there is problem of cooking energy requirement the goal and objectives in terms of total quantum of energy required, form and grade of energy required and which energy sources is required on priority etc. Must be encouraged and assured.

(c) Establishment of Model for the System

Once the problem has been defined and the goals & objectives specified, the next step is to consider how these goals can be met and these objectives can be satisfied within the technical, social and economical setting of energy sources and system. To help this effort, a "model" of the technical, social & economical setting in which the project will be implemented, should be established.

A model is the representation of actual phenomenon process of a system. It should do the following things;

1. Model must explain the system, means which energy sources are being utilized to meet out energy requirements in a particular sector must be cleared through the model.
2. With the help of the model, one can predict the system i. e. energy forecasting is possible from model.
3. Model can help in controlling the system, means proper planning and policy evaluation is possible through model.

It is important to remember that a model should be complete and as accurate as possible & model most have reality and manageability.

Information in the model, which can serve a "baseline", when a rural energy related project is evaluated, should include:

1. Technical Characteristics of system.
2. Environmental & Social descriptions.
3. Economic conditions.

(d) Specification of Criteria of Acceptance

The criteria of acceptance are guidelines against which the rural energy sources/alternatives can be evaluated. Importantly, on one "set of Criteria" will be sufficient for "Judging" the applicability of the proposed alternatives. The criteria of acceptance can be based on:

1. Technical feasibility

It must indicate the feasibility of a particular energy sources or device for a specific purpose technical wise.

2. Economic objective, such as commonly applied criteria to

- Achieves a specified production level of energy at required grade & quality at least cost.
- Maximize benefits and the returns on an investment made in making available a particular energy sources or device.

3. Social and environment considerations

The propose energy system & device must be socially accepted & there should not be any adverse effect on environment.

The criteria of acceptance which reflect the principles of "appropriate technology" are also considered in rural energy problem. These criterions can be require a project to:

- Make "optimal" use of locally available energy sources, which are technically & environmentally sound.
- Make use of, and when appropriate, adopt "traditional" technologies & approaches to problem solving, including exploitation of improved methods to increase the resources.
- Have the potential for being implemented / maintained by the community.
- Have a "reasonable" time frame for the community to take responsibility for the project activities.
- Increase the potential for community "self reliance" in both the short & long terms.

(e) Formulation of Alternatives

At this stage in the planning frame works, it becomes necessary to formulate alternatives, including sources, devices alternatives and alternatives methods of implementation. Because, a "right way" to approach a problem is seldom available, these alternatives must be considered carefully.

(f) Testing of Alternatives

The alternatives formulated must be tested against the criteria of acceptance. In rural energy planning process the formulated alternatives must meet the technical, social, economic & environmental criteria. A particular energy sources or devices or specific activity must be technically feasible, economically viable, socially acceptable and environmentally sound.

(g) Selection of the Best Course of Action

From the above testing, the "best course of action" means selection of best option for satisfying each categories of energy need must be made. Further, the best course of action selected, must be both in terms of a specific project & the methods for implementation and finally integration of the selected opinion into a system, must be made.

(h) Review & Critique

It is important that, once an integrated rural energy planning programme has been implemented, a continual review and critique must be made of the progress of the project. Importantly, all of the people involved in the design, implementation & maintenance of the devices should be included in the review & critique activities. When it becomes apparent from a review and critique that adjustments in the condition of the project are necessary, these adjustments should be made as quickly & efficiently as possible.

Importantly, the information utilized in formulating the model of the technical, social & economic setting of the programme, once again, can be useful in monitoring of the whole plan based on changing technical & socio - economic factors.

2.6 Integrated Energy Planning

Keeping in view the fundamental problem of rural energy to the village in Indian context, it is worth wise to include all energy sources (conventional & non conventional) in holistic manner, so that basic requirement can be fulfilled with locally available environmentally sound sources. This holistic view of energy planning is termed as integrated energy system. This is expected to result in optimum matching of the energy needs with availability. In such a system the emphasis should be on the energy sources which have local decentralized options and are ecologically sound, mostly new and renewable energy sources. Thus, at rural level, the self reliant environmentally sound and need oriented development can be carried out be integration of energy consumption pattern, energy needs, technological options and appropriated energy sources and devices. This type of the system would definitely helpful in the country like India, where more than 70 per cent of population lives in villages. In general, the concept of integrated rural energy system encompasses the following feature:

- The need to promote change from energy development to energy autonomy.
- The need to enlarge the slope from electrification to energisation
- The need to shift from external control to self reliance.
- The need to passive & enhance ecology & maintain nature's cycle and
- To move from conventional non renewable to appropriate non renewable and renewable energy sources.

Several studies have been made which put forth the spectrum of methodology, consideration & objectives which should be kept in view in planning the rural energy supply system. The approach of integrated energy supply system should involve.

- Assessment of resources in the ecosystem.
- Elucidation of current rural energy consumption pattern.
- Translation of these patterns into a set of energy need arranged according to priority.
- Consideration of the feasible technological options, including the tradition ones, of satisfying these energy needs with available resources.
- Selection of "best" option for satisfying each category of need.
- Integration of the selected option into system.
- Mobilization & training of local personal to act as supervisors.
- monitoring of the plan

2.7 Integrated Energy Planning - A Sustainable Path

The demand for energy is derived from the demand for other goods and services, since it is the service that energy provides in production and consumption that determines the type and magnitude of energy demand. Hence, an integrated energy strategy would not treat demand as an inflexible time path in the future, which needs to be met under any circumstances. An enlightened energy strategy, which uses resources optimally, would cover actions and initiatives on both demand as well as supply. It is often less expensive, for instance, to bring about technological changes or, even better, housekeeping measures to reduce demand than to add capacity for enhancing supply to an equivalent level. An optimal strategy would, therefore, use a combination of both demand- as well as supply-side measures. Further, in maintaining a balance and serving national interest, the energy strategy would need to serve the objectives of equity, security, efficiency, and environmental protection.

2.7.1 Equity

The alleviation of poverty is a major challenge for society as a whole and policy-makers at all levels. Energy policies in the past have not paid adequate attention to the special plight of the poor. It is, therefore, essential now to ensure that meeting the needs of the poor is merged into the mainstream of energy policy, such that they are able to pursue sustainable livelihoods. Provision of good quality and adequate energy is central to the objective of removing poverty, in the absence of which society would suffer from serious disorder and dysfunctional conditions. In the context of energy consumption, considerations of equity would have several dimensions. The most important of these is to ensure that adequate energy of good quality is available and accessible to all, such that the poor are able to meet their needs, particularly of cooking fuel and basic lighting. The aspect of quality is important because the use of biomass fuels with low calorific value has serious health implications on account of resultant high levels of indoor air pollution. Clean and good quality fuels, therefore, need to replace the large-scale use of inferior fuels currently prevalent in India.

Globally, about two billion people do not have access to modern energy carriers, such as electricity, and non-traditional cooking fuels like LPG (liquefied petroleum gas); they rely instead on wood, dung, and other biomass. It is estimated that of these two billion people deprived of modern energy forms, around 700 million live in India. The only effective means to lift this deprived section of society out of their current crisis is to ensure that even if financial support is initially provided to them in some measure to purchase and use clean and good quality fuels, over a period of time they must be able to enhance their incomes to afford these fuels through normal market transactions. In this context, energy has to be seen as part of an overall development strategy. Interventions in the energy sector would provide only short-term benefits unless they are combined with matching interventions in other sectors, which lead to the spread of economic opportunities among the poor.

With increases in income, the demand for energy evolves from basic energy end uses, such as cooking and lighting, to mobility, space conditioning and entertainment, all of which need larger quantities and different forms of energy. It is now apparent that keeping specific technologies out of the reach of the poor restricts their abilities and opportunities for economic development. Strategies for meeting the needs of the poor and ensuring their progression to wider and higher forms of energy uses would be well served by policies that make modern technology available to them, even at low income levels. For instance, solar lanterns using photovoltaic technology are as compatible with the ability and need of the poorest of the poor to use them appropriately as they would be among more prosperous sections of society.

2.7.2 *Security*

There are several aspects of energy security that need to be incorporated into an integrated energy strategy. However, recent experience has shown that while physical restrictions in supply may not be a serious threat, it is the economic impact of fluctuating oil prices that is perhaps the most serious issue for consideration in the security of energy supply. Energy security needs to be enhanced through specific short- as well as long-term measures. In the short term, when technologies and capital assets cannot be changed substantially, ensuring energy security would involve means by which efficiency of energy use is achieved through short term measures, such as better housekeeping in industrial units, better maintenance and upkeep of transport vehicles, improved traffic management and various other ways by which wastage of energy is minimized. At the same time, adequate storage of energy, particularly oil and oil products on which India's import dependence is serious, would become an important supply-side option. The long-term challenge of energy security, however, lies essentially in a series of measures related to securing the supply of hydrocarbons as well as in bringing about a shift to greater use of other forms of energy. This would involve conscious decisions often for restructuring the very nature and mix of economic activities in a country. The development of newer energy forms such as CBM (coal-bed methane) and gas hydrates also present long-term options that need to be exploited.

2.7.3 *Efficiency*

In its broadest sense, energy efficiency would encompass all measures that would bring about efficient use of energy re- sources leading to a reduction in the need to produce energy. This would include production, conversion, transmission/transportation, and consumption efficiencies as well as any measures that can be put in place to reduce the level of demand itself—peak or otherwise. In this sense, energy efficiency measures should not only enhance energy security by spreading available energy resources over a longer period of time but also mitigate environmental impacts and reduce the cost of energy delivery.

2.7.4 *Environment*

The environment, as any other factor of production, can become a non-renewable resource if exploited beyond a threshold level. In recognition of the limits that environmental degradation places on economic growth, the United Nations too has been advocating the development of natural resource accounts that would enable a country to account for a depletion of its environmental and natural resource wealth. The process of producing, transporting, and consuming energy has a significant impact on the environment at all stages. While we are yet to internalize the costs of environmental depletion in a systematic manner, several studies have captured the adverse impact of energy use on the environment. Some of the commonly known environmental impacts of the energy cycle include air pollution, land degradation, and deforestation. Environmental impacts of energy use once again vary with the level and pattern of economic development. The environmental impacts of biomass energy are manifested in the depletion of the very resources on which the poor are dependent and in the serious indoor air pollution adversely affecting the health of the particularly vulnerable section within the poor—women and children. As energy needs increase beyond subsistence applications, environmental problems manifest themselves increasingly in the form of rising levels of ambient air pollution leading to environmental health problems at a local level; acid rain at the regional level; and climate change at the global level.

It is interesting to note here that the perceived relative importance of different air pollution problems changes from indoor to local to global as income levels increase from low to high. However, it needs to be recognized that global environmental effects can greatly influence the energy and technology choices of all countries, developed and developing. Also emissions of GHGs (green house gases) in one country can have adverse climate impacts on another country at the other end of the globe.

3

Survey Methodology for Energy Demand and Supply

3.1 Introduction

There is now latest national thinking going on to conserve the known sources of energy such as firewood, cow dung, agricultural waste, kerosene, oil etc. i.e. by limiting their conventional uses and to propagate the use of non conventional energy especially in the rural area. The large scale indiscriminate use of non commercial fuels fully in the rural areas are not only providing to be inefficient but also inadequate to cope with the growing rural energy demand caused by growing population and growing consciousness among rural people for a better way of life. Planned input of energy are, therefore urgently require to protect the environment and to cope with raising rural energy demand. In order to assess the widening gap between the supply and demand of commercial energy and indiscriminate use of non commercial fuels in rural areas, there is a need for various types of statistical or quantified information to be collected and analyzed in an objective manner and presented suitability so as to serve as a sound basis for taking policy decision to make an optimum utilization of commercial energy in a balanced manner. This will require conducting of survey in areas to be covered.

The statistical data may be collected by complete enumeration or by a sample survey. The earlier method is very cumbersome, time consuming, cost ineffective and requires an intensive development of manpower, if domains of study is quite large. Complete enumeration and sample surveys presuppose the existence of a certain minimum of facilities such as funds, professional personnel for planning the survey methodology and supervision of field operations, sufficiently qualified enumerators, sampling frames, machine tabulation equipment, transport and communication facilities etc. These facilities or combinations there of do not always exist to the extent needed for a complete enumeration survey. As such, sample survey is preferred to complete enumeration for ethic type of studies giving due importance to sampling error.

In the integrated energy planning, there is an objective to find out the present consumption pattern of energy for different end uses, shortages being experienced and people's reaction to using alternate sources of energy is also enumerated.

The survey methods determine - incident, distribution and the interrelation among sociological variable and psychological variable. The sociological variable s consists of age, education, income and attitude, behavior, motivation, belief and knowledge about the problem.

3.2 Sampling Design and Methodology

The sampling methodology consists of the following methods:

- By selecting statistical unit at random based on equal probability which is known as random sampling technique.
- By selecting statistical units for investigation, keeping in view the purpose which such units are intended to serve. It is known as purpose sampling.
- By selecting statistical unit first at random and stage by some other method, keeping in view the purpose of such selection, it is termed as Statistical Random Sampling Technique. The method is used in case of item being of heterogeneous character.

In the present survey to study the random and supply of energy in the rural area of state the systematic random stratified sampling method will be used, first of all, the population or universe for the study will be determined so as to give an equal representation to all things of the state and then, a sample size will be selected.

3.3 Type of Survey

A survey is deified as study of large and small population by selecting & studying samples chosen from the populations of discover the relative incidence, distribution & inter - relations of sociological & psychological variables. Survey, probably because survey developed as a separate research activity along with the development & improvement of sampling procedure. The survey researcher is interested in the accurate assessment of the characteristics of the problem. He wants to known foe maple, what are essential activities in which energy is required? How much energy is required for a particular activities etc.? The approach & the technique of survey can be used of any set of objects that can be well defined, survey focuses on people, the vital facts of people and the beliefs, options, attitudes, motivations, behaviors and knowledge about the problems.

Survey can be conveniently classified by the following methods of obtaining in formations:

1. The Interview and Questionnaire Methods

This is principal method of gathering information. Personal interview based on a set of questions related to subject matter is considering critical tool in this type of method. The success of this type of method depends on interviewer, skill and knowledge. Interview schedules are difficult to construct, they are time consuming, relatively costly, but there is no other information gathered in this survey method includes the so called sociological data mentioned previously - sex, maritial status, education, income & other infrastructure available. Such information is indispensable, since it is used in studying the relation among variables & in checking the adequacy of samples. Other kinds of factual information include what respondents know about the energy scenario? What he did in the past for meeting out energy out energy needs? What they are having? & what they intend to do in the future?

2. Panel Survey

The change in behavior of a set of person is determined through this type of method after a certain time gap. In this case a sample of respondents selected & interviewed & then re- interviewed and studies at a later time. This technique is helpful in accessing feed back and further changes required.

3. Telephone Survey

It involve less time and less cost in communicating massage, but it is not face to face method, hence the accuracy of this type of system is less as compared to other method, also it fails to get detailed information. However, in Indian condition, it is not feasible.

4. Mail Questionnaire

A set of questions is mailed individually to know his status on a specific subject. Major disadvantage of this method is lack of response alongwith inability to check the reliability of response. This technique has serious drawback unless it is used in conjunction with other techniques. Responses to mail questionnaire are generally poor Returns of less that 40 to 50% is common.

A questionnaire is a major tool for the success of the survey methodology, whether it is interview method or other.

3.3.1 Tools of study

A survey is designed to collect comprehensive information from all possible sources. A set of schedules / questionnaire are designed for canvassing the respondents, either by personal interview method or by other method such as by sending them by post. The schedules of enquiry for integrated energy planning may broadly cover selected households. The survey schedules should include indentification of households, educational standard of members, annual income of family, use of various energy resources, cost of energy used, methods of its consumption etc. in order to access the demand of energy in the area.

Survey has contributed much to the methodology adopted. Its most important contribution perhaps has been to rigorous sampling procedure & overall design of the specification of the problem.

Survey researchers are a "flow plan' or chart to outline the design & subsequent implementation of survey. The flow plan starts with the objectives of the survey. List each step to be taken & ends with the final report. First the general & specific problem that are to be solved carefully and as completely states as possible. In ours case, energy assessment problem must include its all aspects. One of the investigator's most important job is to specify or clarify the problem. To do this well, he should not expect just to ask problem what they think about energy, although this may be a good way to being if one does not know much about the subject. He should also have specific question to ask that are aimed at various fact of the problem. Such as how much energy & what form of energy is required in particular sector? Each of these questions should be built into the interview schedule.

Some survey researcher even design table for the analysis of data at this point in order to classify the research problem & to guide the construction of interview questions. The data mentioned in table give hints to the respondent about specific problem say for example, how much wood you are using for meeting out cooking energy requirement? Of course this question depends on the number of family members, food habit and types of food etc. But if surveyor know the structure of family & minimum wood requirement per family member than through these figure from the table, he can give hint to the respondent about his answer. It helps in getting answer from the respondent.

The next plan in the flow is the sample and sampling plan. Because sampling is much complex, first the villagers to be sampled & studied must be defined. Are all persons living in the village included? (village leader, those having more resources, those who are educated & socially reputed). Once the village or block is defined, a decision is made to how the sample is to be drawn & how many cases will be drawn. However, in the integrated rural energy planning survey selected households are interviewed.

In the best survey, random samples are used. Because their are high cost & greater difficulties of execution random samples are often by passes for quota samples. In a quota sample "representativeness" is presumably achieved by assigning quotas to interviewers so many men & women, so many farmers and business men and so on. Quota sampling should be avoided in the integrated rural energy planning survey methodology. While it may achieve representativeness it lacks the virtues of random sampling.

The next large step in survey is the construction of the interview schedule & other measuring instruments to be used. This is laborious and difficult business. The main task is to translate the research question into an interview instrument & into any other instruments constructed for the survey. After draft of the interview schedule & other instrument are completed, they are pretested on a small representative sample of the block or village. They are then revised & put in final form.

The steps outlined above constitute the first large part of any survey. Data collection is the second large part. The interviewers are oriented, trained & sent out with complete instruction as whom to interview & how the interview is to be handled. In the best surveys, interviewed are allowed no latitude as to whom to interview. They must interview those individually & only those individuals designed generally by random devices. Some latitude may be allowed in one actual interviewing & use of the schedule but not much. The work of interviewers is also systematically checked in some manner. For example, every tenth interview may be checked by sending another interviewer to the same respondent. Interview scheduled are also studied for signs of spurious answering & reporting.

The third large part of the plan is analytical. The responses to question are coded & tabulated. Coding is term used to describe the translation of question responses & respondent information to specific categories for purposes of analysis. To aid in the coding, content analysis may be used. Content analysis is an objective & quantitative method for assigning types of verbal & other data to categories. Coding can mean the analysis of factual response data and then the assignment of individual to classes or categories or the assigning of categories to individuals.

Tabulation is simply recoding of the number of types of responses in the appropriate categories, after which statistical analysis follows i.e. percentages, averages, relational indices and appropriate test of significance.

The analysis of the data is studies, collected, assimilated & interpreted. Finally, the results of this interpretative process are reported. Survey has a unique advantage among scientific methods, it is often possible to check the validity of survey data. Some of the respondents can be interviewed again and the results of both interviews checked against each other.

Survey method has the advantage of wide scope, a great deal of information can be obtained from a large population. A large population can be studied with much less expense than that incurred by other method. While survey method tend to be more expensive than laboratory & field experiment & field studies for the amount & quality of information, they yield they are economical. Further more existing educational facilities & personnel can be used to reduce the costs of the research.

Survey method information is accurate. However with these advantage go inevitable disadvantage. First, survey information ordinary does not penetrate very deeply below the surface. The scope of the information sought is usually emphasized at the expense of depth. This seems to be a weakness. A second weakness is a practical one. Survey method is demanding of time & money. Sampling and development of good schedules are major operations in this method. Interview requires skill, time & money survey method also requires a good deal of research knowledge & sophistication. The competent survey investigator must know sampling, question & schedule construction, interviewing & analysis of data and other technical aspect of survey. Such knowledge is hard to come by few investigators get this kind & amount of experience.

In all survey methods a set of questions related to subject matter is essentially be formulated before actual dialogue. The questionnaire should have following characteristics: -

1. The set should indicate clear cut conception of variables involved in the system.
2. Minimum cost and time estimate should be incorporated in the set.
3. A prefect questionnaire format should be developed and the sequence of the question should be according to problem.
4. The set should be flexible to accommodate protesting and editing in order to refine the item.
5. The set should be perfectly reliable and valid.

In questionnaire, the following items are consider important:

1. Content 2. Wording 3. Sequence

1. Content

(a) Is the question necessary for clear understanding to the subject matter?

(b) Are several questions needed for subject matter?

(c) Does the respondent have information regarding question?

(d) Does the question need to more accurate or specific?

(e) Is the content biased or loaded in one direction?

2. Working

(a) Have you used difficult or unclear working?

(b) Is the word biased?

(c) Are there objectionable wording?

(d) Can the question better ask directly or indirectly?

3. *Sequence*

(a) Is the question likely to be influenced by the content of proceeding question?

(b) Does the question come too early or too late in the sequence?

(c) Is question lead to the natural way?

Units of Study

The unit of study may be taken as a household, village or a block or district. In the case of integrated rural energy planning, the unit may be a specified number of similar house holds in a village.

Reference Period

The reference period for the collection of primary and secondary data is pre - determined. Generally, a period of three to five years is taken for the collection of secondary data. The reference period for primary data is the data of survey. For integrated energy planning survey, a period of three years will be desirable.

4

Energy Integration

4.1 Introduction

The energy census survey has become the basis for planning & implementation of the village development based on conventional and non conventional sources which are appropriate to the local needs. The continuous and widespread use of non commercial sources of energy in the rural area is resulting in the large scale destruction of the environment. While the rural people get non conventional fuels at almost zero private cost, the social cost for meeting their energy needs has been increasing over the year. Hence the non conventional fuels acquired a special significance in the energy context.

The rural energy problem is multi dimensional & needs to be tackled through a multi dimensional integrated approach which would include augmentation of energy supply resources, technology transfer for optimum & productive use of resources, R & D inputs in particular for developing low cost energy options for the rural poor, education & training for rural users, linkage with various rural development programmes including village & small scale industries programme and adequate institutional & financial support. The multi dimensional approach for making a village self sufficient in energy planning is illustrated in Fig.4.1.

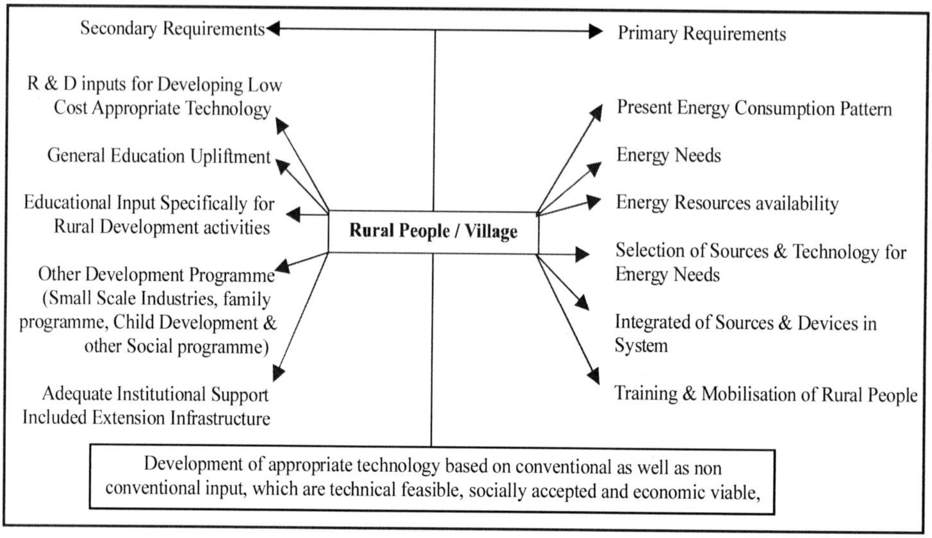

Fig.4.1. Multi Dimensional Approach for Energy Planning for Self Sufficiency

4.2 Integration of Sources and Demand

In order to design a rural energy center, few general principles for the integration of energy sources and devices into a system have to be envolved. The fundamental problem of rural energy planning is to know how to use energy resources optimally in order to meet out energy requirement. At the same time, it is also essential to choose a prefect device having maximum efficiency. In other words, if the combination of energy sources with devices is designated for a particular job as an energy path by which a particular source is utilized with the aid of a device to fulfill an energy task than there are unanswered questions like what is the optimum set of energy paths, and optimum energy flow along each one of this set, to meet the energy needs with the available resources? An immediate solution is required to meet present needs, but it is also essential to anticipate change and to develop solutions which cater to the contours of future needs.

For a given task, it is the maximization of second law efficiency that determines the minimization of fuel consumption in the case of consumable fuels and of capital cost in the case of renewable sources. The maximization of second law efficiency implies that sources and device must be matched to the task. This matching is facilitated by two thumb rules (a) Do not use a higher quality sources that the task deserves, and (b) For the matching source, choose the device which transfer the most useful work / heat.

The maximization of second law efficiency requires the development of low grade energy sources for low grade task and high grade sources for high grade task. Hence, system integration must involve the principles of mixing sources to match task in general, an optimized rural energy system should be based on mix of energy sources *i*, where *i*'s refer to the sources energizing the set of devices that accomplish the required task. These sources *i* may be primary sources in sense that they are inputs to the system (e.g. conventional or non conventional energy sources), or they mat be intermediate sources which are produces inside the system from primary sources (i.e. electricity) or from other intermediate sources (i.e. exhaust heat from an engine driven by conventional or non conventional porch).

4.2.1 Simple integration

One possibility is mix in which all the energy sources are primary sources. If in additional, all the devices are of single sources single task category, then the result is a system with virtually no element of integration. The system is just a juxta position of separate sources, devices and tasks, such a system can be represented by the networking Fig.4.2 from which it can be seen that the path from sources through devices to task are quite separate and unconnected.

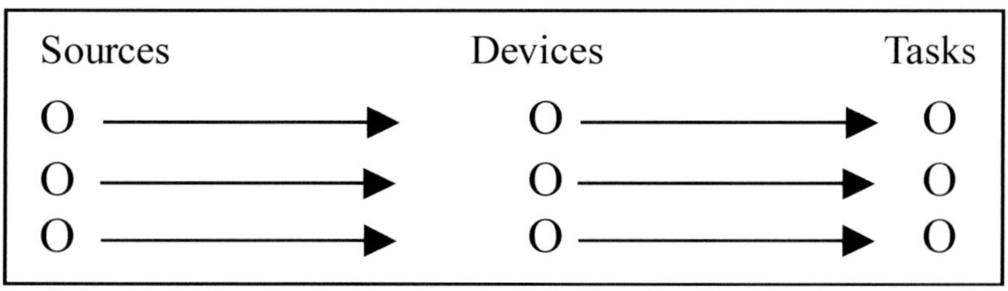

Fig 4.2. Simple Mixing (but no integration)

4.2.2 Principle of cascading

There are other possibilities also available apart from simple integration. For instance, during the course of performing task, device can be produce as waste of lower grade energy. The use of this low grade energy to drive another device which performs a lower grade task illustrates the principle of casing. Fig. 4.3 according to which energy passes from a high quality form to its important final form as ambient temperature heat, it performs a series of task of lower and lower grade. For example, one can think of a series of heat engines each one running on the waste heat from the previous one.

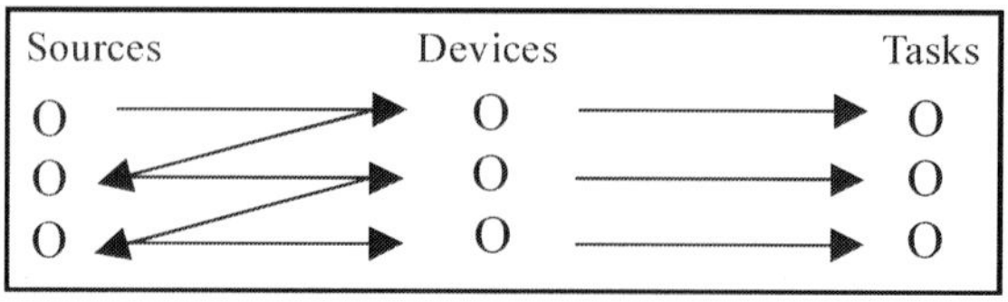

Fig 4.3. Principle of Cascading

4.2.3 Principle of combining

Another approach of integration involves the principle of combining energy sources according to which two or more energy sources act in conjunction to perform a task (Fig. 4.4). Thus, two energy sources can supplemented each other's contribution in heating a fluid e.g. solar preheated water can be used for cooking rice with biogas fuel. The merit of complete the task which is partially accomplished by the lower grade energy.

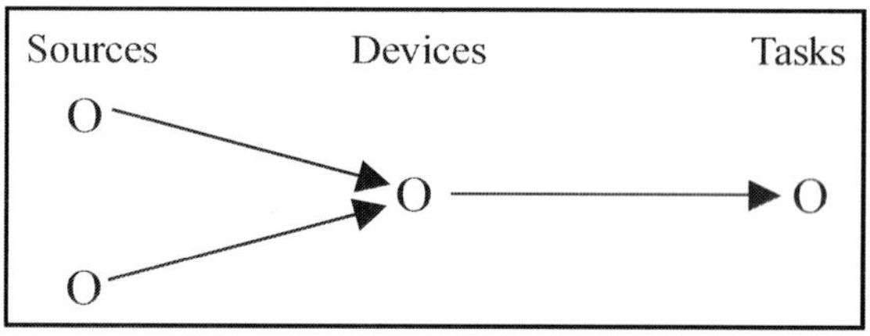

Fig 4.4. Principle of Combining

4.2.4 Principle of cascading cum combining

The two principles of cascading and combining energy sources can be used simultaneously. This requires the waste energy from one device i.e. intermediate sources, being used to supplement the effort of another energy sources in another device or even the same device (Fig. 4.5) for example, exhaust heat from an engine can be used a supplementary heat sources along with a fuel or the waste heat from a sugarcane juice evaporator using bagasse fuel can be used to preheat the juice.

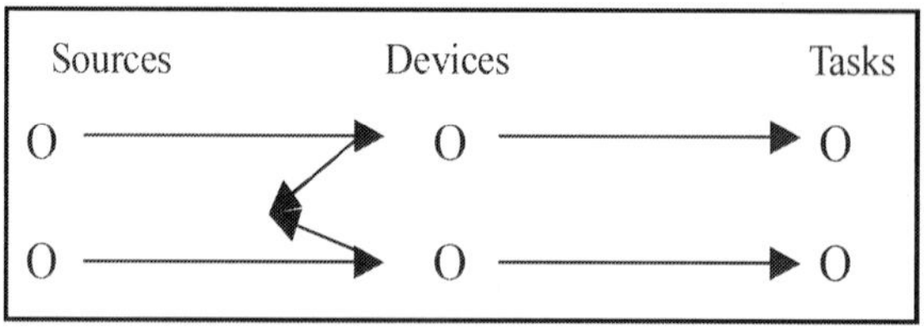

Fig. 4.5. Principle of Cascading cum Combining

4.2.5 Principle of spatial task integration

Instead of introducing integration at the sources end of devices, the integration can also be carried at the task end. This principle of spatial task integration is displayed by hybrid devices which perform more that one task (Fig. 4.6) for example, when the roof of a biogas plant gas holder is made to serve as the absorber of a solar water heater, the result is a hybrid multi task device viz biogas plant cum solar water heater. In fact, the integral nature of the design can be extended one step further by providing a slope to the transparent green house roof of the solar water heater on the biogas plant by collecting the solar radiation distilled water from a gutter built into.

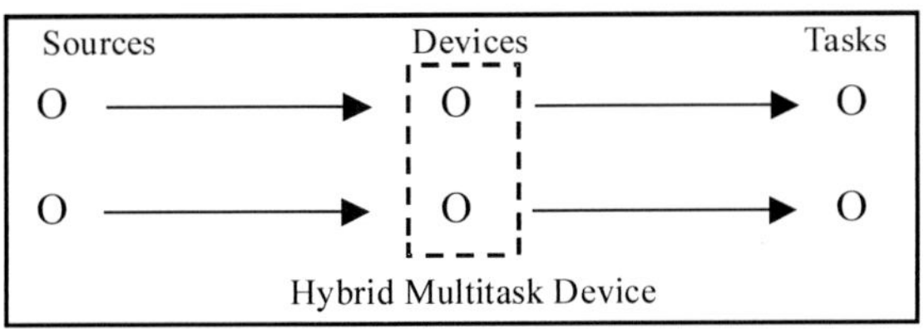

Fig. 4.6. Spatial Task Integration Principle

Time Sharing Principle

The principle of time sharing of device also represented a form of integration (Fig.4.7) for example, an engine or motor can be shared between task which required to be perform at different time or a supply pipe line can be time shared between biogas for cooking and water for domestic consumption.

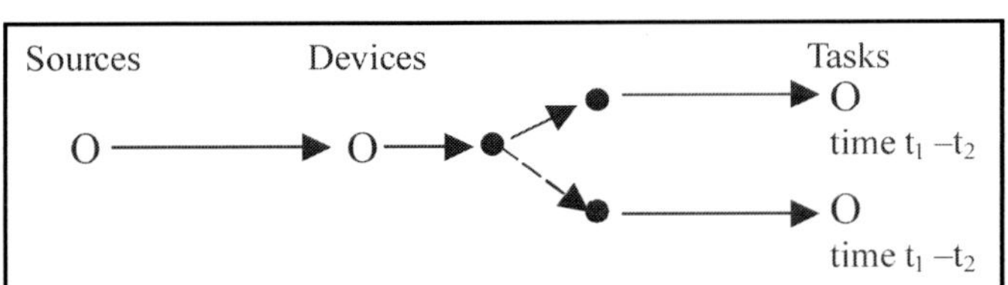

Fig 4.7. Time Sharing

These principle of assembling sources and devices into an energy system may prove useful in meeting energy requirement in different activities.

5

Approaches for Integrated Rural Energy Systems

5.1 Integrated Rural Energy Systems

Basically an integrated rural energy system includes three tier requirements, which are as follows:

- Planning
- Development
- Operation

These three requirements are required to be followed in sequence for getting maximum benefits. However, integrated rural energy concept is a new area of work and not much experience is available in this field as compared to other rural development activities. It is difficult to give any guidelines for planning. One can only recommend approaches and not solutions to planning.

The rural area in developing countries are almost always rich in one or another sources of energy, which could be used separately or in combination with other sources in decentralized small scale units, or even on a household scale. Thus, the problem is one of good planning, management and a strong educational and promotions campaign, rather than the non availability of resources or the lack of technology. Further, after planning, prefect development of device and their effective operation is required.

The purpose of the Integrated Rural Energy Planning project should be clearly defined. It may design either for research and development or demonstration for Popularisation of system, education and training, promotion, health or combination of these.

It is usually better to start with a small size to gain experience and build up confidence before undertaking the large scale project. As defined earlier, one of primary requirement of Integrated Rural Energy Planning is to develop energy demand pattern, energy supply system and optimally match demand with supply.

5.1.1 *Energy demand pattern*

A comprehensive matrix of energy demand in the different sector viz household, community, agriculture, small scale industries should first be establish, including the activities of each sector. For this it may be necessary to conduct survey to collect the data. One can either use available data or a predefined questionnaire to get above requirements.

5.1.2 *Energy supply system*

A matrix should also be established to reflect the rural energy supply system. This should include-

1. Current energy sources.
2. The potential for improving or expanding the existing systems.
3. Possibilities of energy conservation or recycling.
4. New options including new and renewable energy sources.

5.1.3 *Matching demand with supply*

Following consideration should be employed in matching demand with supply for Integrated Rural Energy Planning.

1. Technologies which imply more efficient method of satisfying demand should be used over the less efficient ones. For example the use of firewood in gasifier is preferred to its use in fuel stoves.
2. Preference should be given to energy supply which involves less material or energy waste. Such as solar heaters are preferred over fuel wood or fossil fuels in providing low grade heat in small scale industries.
3. Preference should be given to those technologies which promote recycling. For example preference should be given to biogas over the burning of cow dung as it's of cow dung as its utilization

only as fertilizer, where as biogas provides both energy and fertilizer.

- Biogas effluents can also be used in ponds for fish and algae production

4. Environmental and health aspects consideration should also be included. For example fuel wood utilization should be viewed with it emission. Also biogas and producer gas are preferred for cooking because of their benefit over health and environment impact.
5. The habits of people and the social acceptability of the system should also consideration. Say for example biogas and producer gas may be preferred to solar energy for cooking because that they do not simply changing of habit from indoor to out door.
6. More direct utilization of energy, including simplicity and low cost of application should be consideration while matching demand with supply. For example the use of wind energy in water pumping is preferred to its conversions to electricity and then its use for pumping water.

5.2 Feasibility Study

The feasibility study should be undertaken for each Integrated Rural Energy Planning Project. The study include

- Technical
- Environmental
- Political and economic aspect

The following points are important as far economical feasibility study for matching demand with appropriate supply is concerned.

1. It is extremely difficult, if not impossible to undertake a comprehensive economic analysis, it should include running and capital costs of the various alternatives and also the environmental and health factor. For example impact of reducing emission, switching over from fuel wood to biogas. Here health and environment both are improved.
2. Comprehensive economic analysis should also consider the social impact, many of which are extremely difficult to express in monetary terms. For example:

- Creation of new jobs.
- Improving health and education standards provision of new recreational possibility and their consequent effect on the migration process from rural to urban areas.

3. A simple economic analysis from the consumer point of view has often been found more useful.
4. It may also indicate the foreign currency saving and their possible effect on the balance of trade.
5. While most of economic solutions are usually favoured in urban planning, this is not the rule with rural planning in developing countries.
6. The economic aspect of project or alternative should not be over emphasized at the cost of intangible social and environmental aspects.
7. Availability and reliability of data are very important. If necessary, pre - feasibility study to compile data, review their reliability should be undertaken.
8. New technologies should suit local conditions. One reason for the failure of service project on rural energy or renewable energy in developing countries was that the device or technology was developed and tested in one place and applied in another with different conditions.

The degree of feasibility of a project as a rule of thumb, can be measured by –

- Its dependence on local resources,
- Its involvement of local expertise and potential users, and
- The degree to which the urgent need of the people can be satisfied.

5.3 System Design

Alternative design should be introduced and several options should be considered while Planning Rural Energy Systems. An optimum design which may or may not be a single identified option but could be mixture of several design and options. Villagers should be involved in the selection process. Following are the desirable features of Integrated Rural Energy System design.

1. Simplicity of the system to match the life style of the people.
2. Easy manageability and maintenance of the system.
3. Flexibility of the system including flexibility in replacing equipment or spare part and accepting modification and possible expansion.
4. Optimizing the energy mix. It includes optimal utilization of certain energy sources, particularly new & renewable energy sources. It also includes the rational use of available resources through effective conservation and adequate substitution.
5. Site selection for commissioning of device should be made on the basis of technical suitability, availability of resources, transportation, human resources and social aspects.
6. Subsides should be arranged for individual system like biogas, woodstove, etc. However, problems arise where kerosene, diesel etc. are highly subsidized.

5.4 Implementation

Any action in the rural areas of developing countries should be implemented in stages, starting with those measures expected to provide greatest social efficiency, as in

- Positive contribution to essential requirements.
- Health protection and environment protection.
- Communication and education.

Examples of such system are as follow:

a. Biogas plant.
b. Improved cookstoves.
c. Small scale solar pumps
d. Solar refrigerator for vaccines.
e. Solar television for community education.

The above systems are identified as matured and reliable tools with a very high degree of social efficiency. However, in the implementation of Integrated Rural Energy Planning projects, some of the other considerations are equally important such as:

- Training of local people for operation, repair & maintenances. The training can be conducted through audio visual demonstration and discussion on subject matter.
- Participation of women in planning and implementation of projects. Since women are basically end – users for domestic activities.
- Realistic time schedule and implementation of different devices according to requirements and stages.
- Arrangements for checking the acceptability of the system to the users and for modification, if necessary.
- Effective organization of responsibilities.
- Availability of funds and their periodic release.

The success of Integrated Rural Energy Planning project also depends on the way on which the project is operated. However, effective operation of Integrated Rural Energy Planning project require few important step which are listed as below-

- Problem should be identified and different solution should be discussion before finalizing strategy for implementation.
- The problem can be solved with new and unfamiliar technologies.
- Complementary solution/activities may also be introduced to satisfy further needs that appear in the later stage of implementation/operation.
- It is important that from the very beginning villagers should feel that they are closely involved with the plant.
- Consistency of the system with local culture and traditions. Come time it is necessary to interfere with or even disturb the social structure which must be fully understood. This change can only be slow and possible only if the new technology is very attractive.
- Use of local resources has a pronounced effect on the total cost of the system and on its long term continues operation.
- Project should, whenever possible, pay for themselves.
- Commercial aspect of the project should be given adequate attention.
- Consideration should be given to those trades which might be abolished by the introduction of new technology.

- Satisfying local needs is an important goal of any system / design.
- Training of villagers for operation is equally important. Training courses for local staff are also recommended, before or during the running of the project.
- Maintenance, including availability of specific parts, should be assured.
- Documentation on the project should be provided in the local language, so that beneficiaries would be familiar with the technologies aspect of the devices.
- Regular monitoring and evaluation of the project should be undertaken. Simple installation and processes are desirable.
- Follow-up is important. Through follow-up, one can list modification & other R & D input required for different energy devices & systems. Further, follow-up may also be used to expand activities which have been proved useful and to modify those which have not been so useful.
- Environmental and health aspects of alternative designs should not be overlooked. The ability of the system to integrate other projects for improving health and environmental conditions in the rural areas, such as waste management and recycling projects is an important advantage.
- Each feasible alternative design should be reviewed and evaluated as to the effectiveness of the attempt made to minimize the costs and maximize the benefits.
- Financing for implementation of project should be considered. The finances may be from National, International, local and mixed sources.
- Preference should be given to those systems which satisfy most of the local needs, not only in the area of energy, but in areas like sanitation, education, food, health, water supply and transport.
- An appropriate institutional arrangement has to be set up, starting with decision makers and ending with the villagers, e.g., cooperative, non-governmental organizations etc. are more successful.
- Universities and research institution are usually involved because of their access to decision makers and for their expertise.

- Availability of expertise is indispensable for operation of the system. In some cases, it was found necessary to have expertise in sites for at least five years to ensure the complete diffusion of technology to the local people.

6

Energy Forecasting

6.1 Introduction

Energy planning is possible only with knowledge of past and present energy consumption and future energy demand. There are number of factor which affects the energy consumption pattern significantly. These include energy prices, availability, reliability of supply, uniformity of quality, convenience in use, technical and economic characteristics of energy using equipment & appliances, population growth, income rate of urbanization, educational status, social acceptance, social habit, present knowledge of beneficiary etc.

The energy planning differ in planning sectral wise and planning sources wise. For sector planning purposes, estimates are needed of the short and long run price elasticity for every specific energy resource. Where as in planning sources wise, one must consider its long term availability & its costs.

In general, to cope up future risk and uncertainty, we need to predict the future event. The likely future event has to be given from and content in terms of projected course of variables i.e. the forecasting. Projection is of two types - forward & backward. It is the forward projection of data variable, which is named as forecasting.

6.1.1 The significance of energy demand forecasting

One may ask the question what is the use of demand forecasting? The significance of demand forecasting in the content of policy decision can hardly over emphasized. Demand forecasting is crucial for corporate planning.

In fact, there are three inter-related reason for the importance of accurate energy demand forecasts. The first is that the timely & reasonably reliable availability of energy supplies is vital for the functioning of a modern economy. The second is that expansion of energy supply systems usually many years. And the third is that investments in such system generally are high capital intensive, accounting on average, for some 30 per cent of gross investment in most countries. If forecasts are too low, energy shortage may develop whose cost are usually large multiple of the volume of energy not supplied, but if forecasting are too high, large amount of capital with high opportunity cost might be uselessly tied up for long period of time. Either of these consequences is costly to an economy or more costly from the resources that may have to be warshaled to undertaken detailed & reliable demand studies that could help to avoid such error.

It may be noted that the purpose of forecasting in general is not to provide and exact future data with prefect precision, the purpose is just to bring out the range of possibilities concerning the future under a given set of assumptions. In other words, it is not the "actual future" but the likely future" that we build up through forecasts. Such forecasts do not eliminate, but only help to reduce the degree of risk & uncertainties of the future so that we can undertake operational policy decisions & action following there from. Forecasting is a step towards some sort of an approximation to reality. Without demand forecasting planning will be direction less.

Most people view the world as consisting of a large number of alternatives. Futures research evolved as a way of examining the alternative futures and identifying the most probable. Forecasting is designed to help decision making and planning in the present.

Forecasts empower people because their use implies that we can modify variables now to alter (or be prepared for) the future. A prediction is an invitation to introduce change into a system.

There are several assumptions about forecasting:

1. There is no way to state what the future will be with complete certainty. Regardless of the methods that we use there will always be an element of uncertainty until the forecast horizon has come to pass.

2. There will always be blind spots in forecasts. We cannot, for example, forecast completely new technologies for which there are no existing paradigms.
3. Providing forecasts to policy-makers will help them formulate social policy. The new social policy, in turn, will affect the future, thus changing the accuracy of the forecast.

6.2 Technique of Demand Forecasting

Based on term, demand forecasting can be classified as short - term forecasting & long term forecasting. There are specific techniques which fall under each of these broad methods.

In general, the energy demand forecasting is difficult on account of various factors, some of them are as follows.

1. The rate and pattern of economic activities is changing day by day.
2. The technology trend of production of goods and energy is changing day by day.
3. The difference in energy efficiency in various sector and substitution of present energy sources (non commercial energy) by commercial energy.

Some of the above factors on which demand for energy depends, are qualified and some are intangible factors which cannot be qualified easily and directly. Therefore, the choice of energy demand forecasting technique, the choice of energy demand forecasting technique depends on the following factors:

1. The precision of estimate required.
2. The time horizon for which the demand is to be estimated.
3. The accuracy and period of the time series or cross section data avialble.
4. Computing facilities available.

The commonly used techniques for energy demand forecasting are:

1. Trend extrapolation of curve fitting.
2. Input - output analysis (macroeconomic analysis).

3. Econometric multiple correlation forecasting (correlation and regression analysis).
4. Surveys.

6.2.1 *Trend analysis*

Trend analysis is the most commonly used approach for short term as well as quick estimate, which is based on extrapolation of post growth trends, assuming that there will be little change in the growth pattern, determinants of demand such as income, price, consumer tastes etc. These trends are usually estimate by least square fit of past consumption data or by some similar statistical methodology. Based on the availability of data, they may be estimated on a national basis for sectoral wise (i.e. household, industry, transportation, agriculture etc.) or for source wise (i.e. coal, oil, natural gas etc.) or by both.

The main advantage of this approach is its simplicity and forecast can be made on whatever data are available. But, in this type of analysis no attempt is made to explain why certain consumption trends were establish in the past. In addition, it seems to be inaccurate on account of fast change in energy prices.

In this analysis, trend is worked out by fitting a trend equation to time series data with aid of an estimation method. The equation may be either liner or non linear form. Some of the trend equations are: -

(a) Linear Regression

A simple forecasting method that calculates a straight line. By its nature, the straight line it produces suggests that it is best suited to data that is expected to change by the same absolute amount in each time period. The mathematical equation shows that the variable y varies by a constant a and increasing (or decreasing) over time (denoted by t) by factor b.

$$y_t = a + bt$$

(b) Exponential Function

This method uses an increasing or decreasing curve rather than the straight line of the Linear Regression method. An exponential method is useful when it is known that there is, or has been, increasing growth or decline in past periods.

$$y_t = ab^t$$

(c) Power Function

This method is similar to Exponential Function, but produces a forecast curve that increases or decreases at a different rate.

$$y_t = at^b$$

(d) Logarithmic Function

This method is similar to Exponential Function, but uses an alternate logarithmic model.

$$y_t = a + b \log(t)$$

(e) Parabola Function

This method attempts to fit a 'Parabolic' (second order polynomial) curve.

$$y_t = a + bt + ct^2$$

6.2.2 Input - output analysis (macro economic analysis)

For long term estimated & for more accurate estimates technique like input - output may be used. The technique which account for the inter - sectoral relations & intermediated products (input - output) is used for consistency but require much more data for all sectors in the region / nation.

Basically macro economic & input - output energy model connect a number of energy variables with how total energy use & price result from a driving force on the demand side (say, consumer income) & a constant on the supply side (say, the cost of production of energy). In detailed model, there is cleared difference between various end uses of energy (energy sector) and energy sources. This is generally estimated on the basis of empirical information, and the major source for empirical information are technological information relating to past & future production, utilization possibilities and econometric estimates of past demand, supply and other economic behavior relations. Much of the technical information consists of physical quantities of inputs and outputs.

The major advantage of such comprehensive modeling based on input and output are that they internally consistent estimates of future demand and supply relationships and they make it possible to test the effect of various assumptions. However, for such model analysis a quite detailed and reliable set of statistical data are required.

6.2.3 Econometric multiple correlation forecasting (regression and correlation analysis)

Econometric forecasting technique are usually more sophisticated and in theory, hold out the the promise of greater forecast accuracy. Basically in such approach, firstly past energy demand is correlated with other variable such as prices & income and then future energy demands are related to the predicted growth of these other variables. The one of the problem in such estimates is the data availability and even more frequently such are estimates are subject to substantial error.

(a) Regression and correlation analysis

Regression analysis involves identifying the relationship between a dependent variable and one or more independent variables. A model of the relationship is hypothesized and estimates of the parameter values are used to develop an estimated regression equation. Various tests are then employed to determine if the model is satisfactory. If the model is deemed satisfactory, the estimated regression equation can be used to predict the value of the dependent variable given values for the independent variables.

Regression model

In simple linear regression, the model used to describe the relationship between a single dependent variable y and a single independent variable x is $y = a_0 + a_1x + k$. a_0 and a_1 are referred to as the model parameters, and is a probabilistic error term that accounts for the variability in y that cannot be explained by the linear relationship with x. If the error term were not present, the model would be deterministic; in that case, knowledge of the value of x would be sufficient to determine the value of y.

Least squares method

Either a simple or multiple regression model is initially posed as a hypothesis concerning the relationship among the dependent and independent variables. The least squares method is the most widely used procedure for developing estimates of the model parameters.

(b) Correlation

Correlation and regression analysis are related in the sense that both deal with relationships among variables. The correlation coefficient is a measure of linear association between two variables. Values of the correlation coefficient are always between -1 and +1. A correlation coefficient of +1 indicates that two variables are perfectly related in a positive linear sense, a correlation coefficient of -1 indicates that two variables are perfectly related in a negative linear sense, and a correlation coefficient of 0 indicates that there is no linear relationship between the two variables. For simple linear regression, the sample correlation coefficient is the square root of the coefficient of determination, with the sign of the correlation coefficient being the same as the sign of b1, the coefficient of x1 in the estimated regression equation.

Neither regression nor correlation analyses can be interpreted as establishing cause-and-effect relationships. They can indicate only how or to what extent variables are associated with each other. The correlation coefficient measures only the degree of linear association between two variables. Any conclusions about a cause-and-effect relationship must be based on the judgment of the analyst.

6.2.4 Survey methods

The survey method provides a direct & reliable tool of demand analysis & forecasting. Viewing to the limitation of other forecasting methodologies, survey method is quite promising source of estimation. Practically, this method consists of set of questions that are put to energy uses in order to measure & record their present energy consumption & future consumption plan. In this context, some of the basic questions to be asked to users are listed as follow:

1. How much energy (of each type) do you use month wise or year wise?
2. What do you use it for?
3. How much do you pay per unit energy used (by energy type)?
4. What do you produce or sell & what is the value & value added of each major product line?
5. Can you identify specific energy uses with specific output?
6. What is your net income?
7. What are your future expansion plans in energy consuming field?

However, the survey method is affected by different problem such as:

1. They require substantial amount of time.
2. They can be costly.
3. They require skilled interviewers.
4. Energy uses may be unable to provide the information asked because they themselves do not know.
5. Energy users may be unwilling to provide the information for various social & religious reasons.
6. Energy users may willingly or unknowingly give inaccurate answers.
7. Future energy use plans may be vogue or too optimistic/ pessimistic.

6.2.5 *End use*

One of the technique used by our planning bodies for demand forecasting of energy is "end use" technique which is more an accounting technique.

6.3 Energy Demands Forecasting Technique V/s General Forecasting Technique

a. Aim

Not connected with 'what it' factors, feedback loop and regulatory factors

Methods: Qualitative technique are recommended. These are

A1 - Life cycle analysis

A2 - Delphi study

A3 - Seth - hanstha method

A4 - Idea engineering consenses

A5 - Energy surveys.

b. Aim

Less concerned with understanding of inter-connections and casulities of factors.

Method: "Time Series" methods are recommended for use, among these most common are:

B1 - Extrapolation

(a) Simple extrapolation

(b) Curve fitting with judgment modification.

(c) Systematic or statistical curve fitting

(d) Tend extrapolation.

B2 - Smoothing

(a) Average simple (mean)

(b) Moving average

(c) Exponential smoothing

(d) Adaptive exponential smoothing

(e) Double or higher order moving average

B3 - Decomposition methods

B4 - Autoregressive methods

B5 - Filtering methods - Kalman filtering

All the above technique fall under "Time Series" or "Quantitative Technique".

c. Aim

Consider and include the casualties and inter connection.

Methods: The other quantitative technique which are not based on more time - trends are correlative but also on inter relation among various factor, dependence on multiple factors and casualties.

These are:

C1 - Input - output technique or inter - industries technique

C2 - Econometric model or simultaneous equations.

Method: Multi equation simulation models.

d. Aim

More concerned with 'what if' concerns and feed back.

Method:

D1 - Cross impact metrics analysis

D2 - System - dynamic techniques

D3 - Simulation.

6.4 Demand and Supply Scenario: - Source Wise

In the recent years, India's energy consumption has been increasing at one of the fastest rates in the world due to population growth and economic development. Primary commercial energy demand grew at the rate of six per cent between 1981 and 2001 (Planning Commission 2002). India ranks fifth in the world in terms of primary energy consumption, accounting for about 3.5 per cent of the world commercial energy demand in the year 2003. Despite the overall increase in energy demand, per capita energy consumption in India is still very low compared to other developing countries.

India is well-endowed with both exhaustible and renewable energy resources. Coal, oil, and natural gas are the three primary commercial energy sources. India's energy policy, till the end of the 1980s, was mainly based on availability of indigenous resources. Coal was by far the largest source of energy.

Despite increasing dependency on commercial fuels, a sizeable quantum of energy requirements (40 per cent of total energy requirement), especially in the rural household sector, is met by non-commercial energy sources, which include fuelwood, crop residue, and animal waste, including human and draught animal power. However, other forms of commercial energy of a much higher quality and efficiency are steadily replacing the traditional energy resources being consumed in the rural sector.

Resource augmentation and growth in energy supply has not kept pace with increasing demand and, therefore, India continues to face serious energy shortages. This has led to increased reliance on imports to meet the energy demand.

6.4.1 Coal

India now ranks third amongst the coal producing countries in the world. Being the most abundant fossil fuel in India till date, it continues to be one of the most important sources for meeting the domestic energy needs. It accounts for 55 per cent of the country's total energy supplies.

In order to analyse and forecast the demand for coal it is useful to divide it into several sub groups as Anthracite, Bituminous coal, Sub-bituminous coal, Brown coal / lignite, Peat and Coke. Because carboniferous or coal resources occur in a wide variety of physical characteristics or local quantities, the various subgroups presents a simplified division in terms of calorific values only. From a consumption point of view, a more useful division might be one of that divides existing coal into cooking coal, steam coal & lignite. A major problem contributing the wider utilization of coal in transportation cost.

The potential, future demand will depend on the projected industrial & utility needs for processing heat & high pressure steam and cost of delivered coal relative to fuel oil, natural gas & LPG. The rate at which new plants capable of using coal will be built.

Through sustained increase in investment, production of coal increased from about 70 MT (million tonnes) in early 1970s to 382 MT in 2004/05. Most of the coal production in India comes from open pit mines contributing to over 81 per cent of the total production while underground mining accounts for rest of the national output. Despite this increase in production, the existing demand exceeds the supply. India currently faces coal shortage of 23.96 MT. This shortage is likely to be met through imports mainly by steel, power, and cement sector.

The development of core infrastructure sectors like power, steel, and cement are dependent on coal. About 75 per cent of the coal in the country is consumed in the power sector.

6.4.2 Electric power

The electic power sector is one of the most important elements within the energy frame work. In India electricity in domestic application is being used in Air conditioning (30 per cent), Refrigeration (25 per cent), Water Heating (15 per cent), Cooking (7 per cent), Grinding (5 per cent), Entertainment (5 per cent) and Washing (3 per cent). The demand analyses and load forecast for this source have a better foundation. However, forecast error, on the other hand, are usually more serious because of long lead time for new equipment installation, their initial capital cost & the difficulty of replacing electricity by other energy sources, except for cooking, heating and domestic lighting purpose. But even for these uses different appliance are needed. Forecasts that turn out to be low may lead to costly, short - term solutions for alleviating capacity shortages, such as the installation of high cost generating capacity plants (diesel or gas turbines) or to rationing or forced supply interruptions. Local forecasting, therefore play a critical part in the electric systems planning process.

A knowledge of load at the disaggregate level is required because the demand characteristics vary by type of consumer, by geographical area and over time & therefore the properties of the aggregate demand (i.e. at the system level) may be quite different from the characteristics of the individual loads. Whenever possible, survey approaches should be used to forecast the demand of large users. This may account for a substantial percentage of the total load. In all, demand forecasting is a dynamic process in the sense that the procedure must be repeated often in light of improved information & technique of alaysis.

In absolute terms, the installed power capacity has increased from only 1713 MW (megawatts) as on 31 December 1950 to 1, 18, 419 MW as on March 2005. The all India gross electricity generation, excluding that from the captive generating plants, was 5107 GWh (gigawatt-hours) in 1950 and increased to 5, 65, 102 GWh in 2003/04.

Energy requirement increased from 390 BkWh (billion kilowatt-hours) during 1995/96 to 591 BkWh (energy) by the year 2004/05, and peak demand increased from 61 GW (gigawatts) to 88 GW over the same time period. The country experienced energy shortage of 7.3 per cent and peak shortage of 11.7 per cent during 2003/04. Though, the growth in electricity consumption over the past decade has been slower than the GDP's growth, this increase could be due to high growth of the service sector and efficient use of electricity. Per capita electricity consumption rose from merely 15.6 kWh (kilowatt-hours) in 1950 to 592 kWh in 2003/04. However, it is a matter of concern that per capita consumption of electricity is among the lowest in the world. Moreover, poor quality of power supply and frequent power cuts and shortages impose a heavy burden on India's fast-growing trade and industry.

6.4.3 Natural gas

The analysis & forecasting of natural gas demand raise some special issue. While the measurement of existing or past consumption is relatively simple because supply networks are closed systems with metering both ends, forecasting future demand is not, because of the relative ease of substitution by other energy sources.

A second problem for the demand analysis of gas is that supply requires heavy front end investment & that these investments are subject to substantial economics of scale. Another problem that increase fixed cost result from the fluctuating nature of demand. However, demand forecasts for natural gas can be conveniently divided into four groups as shown in Fig. 6.1.

India's consumption of natural gas has risen faster than any other fuel in the recent years. Natural gas demand has been growing at the rate of about 6.5 per cent during the last 10 years. Industries such as power generation, fertilizer, and petrochemical production are shifting towards natural gas. India's natural gas consumption has been met entirely through domestic production in the past. However, in the last 4/5 years, there has been a huge unmet demand of natural gas in the country, mainly required for the core sectors of the economy. To bridge this gap, apart from encouraging domestic production, the import of LNG (liquefied natural gas) is being considered as one of the possible solutions for India's expected gas shortages. Several LNG terminals have been planned in the country. Two LNG terminals have already been commissioned: (1) Petronet LNG Terminal of 5 MTPA (million tonnes per annum) at Dahej and (2) LNG import terminal at Hazira. In addition, an in-principle agreement has been reached with Iran for import of 5 MTPA of LNG.

6.4.4 Petroleum products

Petroleum product on account of their relative cleanness, convenience in use, easy transportability has made them the preferred fuel. However, the demand for them is not entirely inelastic. There are five major type of petroleum substitute available (Fig. 6.2).

Because of the potential availability of the substitute, the demand for petroleum based energy resources can be completely price inelastic, provided price changes are large. The future demand of petroleum products are affected by specific variables such as energy use & price, petroleum product & alternative energy resource prices, potential investments in alternative energy supply system, potential technological change, the cost & feasibility of energy conservation measure. In other words, the demand projections should be based on estimates of the technological socio economic & managerial feasibility of utilizing either petroleum based or alternative energy sources, the potential for reducing petroleum energy input per unit of output or unit of consumption and the likely effects of government policies related to energy using activities.

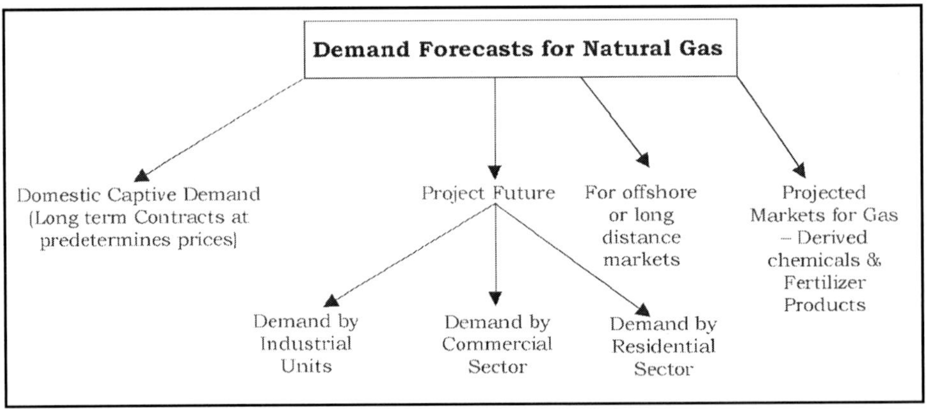

Fig 6.1 Demand Forecasts for Natural Gas

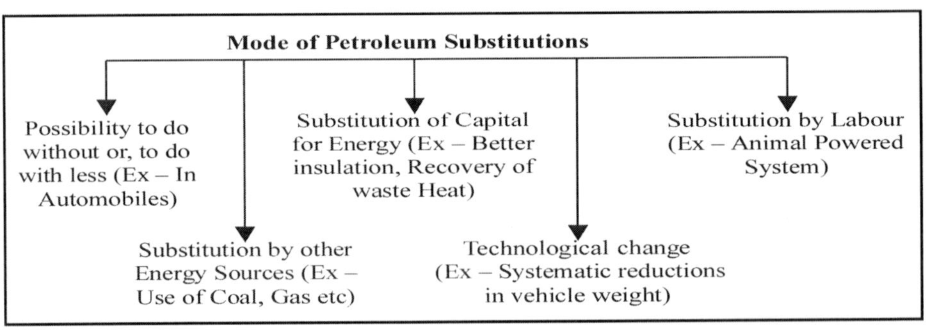

Fig 6.2. Mode of Petroleum Substitution

The latest estimates indicate that India has around 0.4 per cent of the world's proven reserves of crude oil. The production of crude oil in the country has increased from 6.82 MT in 1970/71 to 33.38 MT in 2003/04. The production of natural gas increased from 1.4 BCM (billion cubic metres) to 31.96 BCM during the same period. The quantity of crude oil imported increased from 11.66 MT during 1970/71 to 81 MT by 2003/04. Besides, imports of other petroleum products increased from 1 MT to 7.3 MT during the same period. The exports of petroleum products went up from around 0.5 MT during 1970/71 to 14 MT by 2003/04. The refining capacity as on 1 April 2004 was 125.97 MTPA (million tonnes per annum). The production of petroleum products increased from 5.7 MT during 1970/71 to 110 MT in 2003/04.

6.4.5 Renewable energy sources

In present energy scenario, where fossil fuel resources are depleting and their prices are also increasing manifold, the con conventional or renewable energy sources are getting importance. These sources in addition to lessening the burden on fossil fuels can also help in bridging gas between demand & supply to power sector. Presently devices to harness the alternate energy sources solar, biomass, wind & hydro etc. are available the centralized electrical power generation system is no doubt by far the most versatile in nature, but it is not quite suited to meet the important energy requirement of cooking & agriculture in villages from the optimum point of utilization. To meet the decentralized energy requirement for energy consumption & also for best utilization of available resources, unconventional energy sources need to be used in integrated manner.

There is a significant potential in India for generation of power from renewable energy sources— small hydro, biomass and solar energy. The country has an estimated SHP (small-hydro power) potential of about 15 000 MW. Installed combined electricity generation capacity of hydro and wind has increased from 19 194 MW in 1991/92 to 31 995 MW in 2003/04, with a compound growth rate of 4.35 per cent during this period (MoF 2005). Other renewable energy technologies, including solar photovoltaic, solar thermal, small hydro, and biomass power are also spreading. Greater reliance on renewable energy sources offers enormous economic, social and environmental benefits. The potential for power production from captive and field-based biomass resources, using technologies for distributed power generation, is currently assessed at 19,500 MW including 3500 MW of exportable surplus power from bagasse-based cogeneration in sugar mills.

It is quite apparent that coal will continue to be the predominant form of energy in future. However, imports of petroleum and gas would continue to increase substantially in absolute terms, involving a large energy import bill. There is therefore, an urgent need to conserve energy and reduce energy requirements by demand-side management and by adopting more efficient technologies in all sectors.

Taking into account all the end use products and estimate of energy required (to produce) per unit of these products, the total bill of energy required is estimated. In this technique the intermediate products which are never used as end product and energy required to produce these is not account for. However, this used along with econometric technique may produce acceptable forecasting of demand for energy.

7

Modelling for Rural Energy Planning

7.1 Introduction

Energy has become a critical input for the economic development of the India because of the wide spread scarcity of oil and electricity and other commercial form of energy.

The rural areas are often the worst affected because of these shortages and the rural population continues to depend on the inefficient non-commercial energy from like fire wood, cow dung, human and animal for meeting their energy requirement for subsistence, production and devolvement needs.

The rural energy crises or "The energy crisis" has been getting increased attention from the policy makers to tackle the problem in a planned and integrated manner.

The rural energy sector is concerned with two major planning issues.

(i) Provision of energy for meeting substance needs for cooking, heating and lighting of the rural areas of the country. This problem is also directly linked to the problem of deforestation and environmental destruction being caused in the rural area because of indiscriminate and wide spread use of non conventional energy sources.

(ii) Meeting energy requirements for production activities which would create employment and productivity in rural areas.

In the present time of energy crisis, Rural Energy Planners must assure the supply of energy to people living in rural areas in order to obtain a homogenous development of the country. India has made a noticeable technological advantage of the country. India has made conventional and non conventional sources. The availability of energy from different sources depends on region, resource endowment and socio - economic characteristics of region. In order to have an assured energy supply to rural people for subsistence needs and production and development activities, an integrated energy planning is required for optimum combination of all type of energy sources. A methodology or technique must be developed for finding optimum combination of different energy sources for an area or a region. The computer modeling is a technique for all type of planners. For energy planners it will provide an optimum combination of different energy sources on the basis of energy requirements of the area, investment plan for energy and socio-economic aspects.

Further more, researcher, educators and policy makers concerned with energy planning required a variety of information source and service in the conduction of their tasks. One of the most important of the service information is computer databases. Many information based computer databases evolved through the utilization of computers in collecting information and preparing this material for type setting. Soon it was discovered that, with appropriate computer programs, information searches easily could be obtained in much less time, than a manual search of the printed publication. As a result, computer searches for information began to be used widely in the mid - 1970's and they have continued to be popular.

7.2 Energy Modelling

A model is a representatives of a system. The energy model must represent both the demand and supply relationships for energy. Either the supply or the demand side may receive greater attention in any given model. Thus, in econometric models process analysis, mainly technical relationships are used for detailed analysis flows on the production or supply side. Similarly, econometric market models that focus on the economic behavioral reactions of consumers to price, income and other variables are commonly used for modeling the demand side.

Although the development of energy modules began in the early sixties, yet it was the growing awareness of the energy problems originating from the first oil crises in 1973 that forced an explosion in the development of energy models. The energy models developed in the sixties focused mainly upon the supply and demand of single energy from such as electricity, oil or natural gas. For example, petroleum companies have developed and applied large allocation models in order to deal with the complex problem of optimal allocation and routing of crude oil products between different oil sources, refineries and demand centre as well as to optimize the production process in the refinery.

Another example of such model application related to the analysis of electric utility operations and or system expansion planning. Several models using the optimization or simulation approach have been developed and are used to evaluate the optimal expansion strategy of the power plant system required to satisfy an increasing electricity demand.

Similar to the energy models mentioned earlier, energy demand in energy supply system models is usually treated exogenously. Thus energy supply system models do not allows for demand adjustments due to higher energy price or changes in GDP growth (caused by rising energy cost and limited energy supplies) or due to changes in the socio - economic structure and the consumption behaviour.

Therefore, mainly to capture the energy economic interface more appropriate, energy economic models were developed in the past. Such models link the energy sector with the rest of the economy.

7.2.1 Classification of planning models

Depending on the objectives of the planning problem and the nature of requirement desired, an appropriate energy model can be chosen from one of the groups presented in Table 7.1. For each category of models, the most common methodologies are listed in the table.

(a) Energy Demand Models

These model estimates the sectoral energy demand either in final energy demand or in useful energy demand or in energy service demand. Among the energy demand model, the simulation method and econometric techniques are most commonly employed.

Table 7.1 Classification of Energy Planning Models.

S.No.	Energy Planning Models	Purpose	Methodology
1.	Energy - Demand Model	Analysis and assessment of demand variables and demand structures	Econometric, Engineering process analysis approach
2.	Energy Supply Model	Indentification of fuel and technology mix	Simulation Optimization
3.	Macro Economic Model	Analysis of impact and feedback between energy sector and the rest of the economy	Simulation Optimization Econometric, Input - Output System Dynamics

(b) Energy Supply Model

These focus either on single energy form and sector (sectoral supply model) or the energy system as a whole (energy supply system models). A major criticism of a single fuel model is that they treat the development of the fuel in question in isolation from the rest of the energy and economy. Energy supply system models represent the energy supply system by an orientated network and consider entire energy system as opposed to a single sector / fuel.

(c) Energy - Economic Model

These were developed explicitly for linkages and *interrelationship* between the energy sector and the economy in terms of capital, layout, material and other requirements. There are integrated energy economy models which explicitly describe the interaction within a single network of equations or model sets which consist of a separate economy module and an energy module.

Recent developments in energy modeling concentrate on the development of integrated set of models, which may serve a variety of clanning tasks. The links between the different modules are performed by a data management system.

As far as methodology for estimation of each category of model is concerned, it falls into three broad categories, these are:

(a) Optimization
(b) Simulations
(c) Econometric or Input /Output

In optimization methodology model, in which some from of objective function is optimized subjected to physical, economic and policy constraints for example, minimization of the cost of meeting total energy needs given the unavailability of domestic oil resources. These methodologies are unavailability of domestic oil resources. These methodologies are usually of the mathematical programming type with linear as well as quadratic objective functions and constraint. Optimal control models seek to provide a continuous path over time of the control (or policy) variables that will yield the optimal outcomes.

In simulation methodology alternative scenarios are explored in a self consistent frame work, subject to constraints. In this type of methodology the policy maker has to choose from among the alternative simulated scenarios. This also allows more latitude with respect to judgment and incorporation with considerations which cannot be qualified in the model. Such a situation which may be more relevant in developing countries. The simulation methodology is often econometric in nature where technical, economic and behavioral equations represent relationship among different variables in the econometric equations are estimated and the models "turned" on the basis of past data. The emphasis here is on consistency. Whatever the scenario selected, the evaluation toward the final state is constrained by all the variables having to grow in a mutually consistent manner according to prespecified relationships generally established on the basis of historical data.

Input output methodology, energy economic relationships are captured by a matrix of coefficients representing the quantities of different primary factors and other inputs required to produce are unit of output in various sectors. Input - output methodology may be used in both simulation and optimization type framework. These are particularly useful in modeling the complex linkage between the energy sector and the rest of the economy. Another type of methodology, the system dynamics methodology is especially formulated to incorporate the effects of feedback, lagged variables and dynamic changes in rates of charge.

7.3 Important Energy Planning Issues for India

In our country, consumption of non commercial energy forms is quite high. However, the trend in the past few years shows that the share is fast decreasing and commercial energy sources are taking its place. This has two fold implications: First is need for adequate representation of non commercial energy sources in modeling the energy system, second is to analytically evaluate how decreasing share of traditional fuels will can be substituted by non conventional or renewable resources rather than the commercial resources.

The other feature particular to India are low price and elasticity of demand resulting in a slower transition rate, and high energy GDP elasticity, compared to most industrialized countries. These are important issue in consideration of penetration period of new technologies and demand projections.

While considering the supply side, a large potential for renewable energy resources like hydroelectric and solar is a positive feature, while rapid rate of deforestation and low efficiency of use on non commercial fuels are the negative features.

The institutional aspect of importance in modeling is, the energy sector is almost totally government owned in India fuel–price are administrated prices.

7.3.1 Modelling of non conventional and renewable energy sources

The share of non commercial energy sources in the energy supply system in India has been high. However, this share is steadily decreasing due to non availability of traditional fuels like fuelwood, agricultural waste etc. If this share is replaced by commercial fuels, demand rate of commercial fuels, already high, can phenomenally become high, and hence would difficult to be matched by increase in its supply. The preferred option would be that renewable / non conventional sources like solar energy, bio energy, wind energy and other sources take the place of decreasing share of traditional fuels and this option could be analytically examined with the help of suitable models.

Till now non commercial fuels have not been adequately represented in the energy models for two reasons:

1. Most non commercial fuels are essential free of cost, if the time and effort of collecting them is not counted and do not enter

market forces. Their inclusion models, essentially quantitative and economic in nature are difficult.

2. Availability and pattern of use of these fuels are heterogeneous in nature (agricultural residues, dung).

Their homogenous representation required in many energy models is difficult. Non conventional energy sources have also the same heterogeneous nature, potential for wind, micro hydro and solar energy being uneven in different regions. The conversion technologies associated however, have measurable economic costs (e.g. windmill, photovoltaic) which can be well represented in economic models. If regional model suitable for decentralized energy planning are developed, options for increasing the share of non-conventional energy sources replacing non-commercial and some commercial fuels can be analytically examined. This could be helpful in determining pricing / subsidy policy.

7.3.2 Macro - Economic model development

In fact, in present situation, the economic and energy system are inter link with each other. Thus energy planning is not only the goal in itself but there is a need for developing a tool for effecting desired economic growth in a country / region. For this purpose it is important to known how a specific energy system-existing or planned would affect the economic growth and what demands a specific pattern of economic growth make on the energy system of a nation.

It is proposed to develop macro economic model for this purpose which would have suitable link with available modules like, supply system model or demand models. This would to fill in the gap of a model, currently absent in the system of models.

7.3.3 Energy and environment

Environment degradation is a serious problem in present context. A number of environmental issues which concern society are associated with energy production technologies, end-uses or recovery of primary resources. The concern and public awareness regarding these issues is increasing with time. Though some of existing model incorporate environmental aspects in it, other models need modifications to include this aspect in them. Our aim is to modify the existing model to include important environmental concerns such as forest damages, CO_2 increase and rapid deforestation and smoke from open cook stoves.

7.4 Approach for Computer Modelling

The utilization of computer in energy for rural areas is an important component of the new centrally sponsored scheme which was started in 1986-87. Under this component, a computer modeling project for rural energy sector has been taken up for preparation of investment plans for meeting rural energy needs in the most cost effective manner at the National and State levels.

In fact, mathematical representation of a system is called mathematical model of the system. The mathematical model when solved with the help of computers are called computer model. At our country level, a computer model on the main frame of the planning commission computer has been developed and tested by the Rural Energy Division, Planning Commission, Govt of India. This computer model has been successfully validated on real block data for various eco-regions of the country. This computer model provides a quantitative frame for determining the optimal mix on area wise basis and linking this areas mix with total energy requirements for the rural sector. The model utilized mixed integer programming technique to prepare such optimal plans.

Due to it high cost the mainframe computer does not become viable for large scale decentralized planning. Personal computer however is now readily available and due to their low cost are ideal for utilization in rural energy planning at macro level. The main frame micro computer has been therefore adapted during 1987-88 on personal computer environment with the assistance of the National Productivity Council (NPC).

7.4.1 Modelling of an energy system

A mathematical model for an energy system must consider the following factors while development.

1. Cost of energy system (installation and operationing cost)
2. Demand of energy
3. Quality of energy (grade or temperature)
4. Performance of energy sources (production, conversion and utilization efficiency)
5. Adaptability for charge.
6. Conservation of energy and material.

7. Safety of system
8. Environmental consideration.

A step by step listing is though to be helpful in understanding the system. The following procedure is widely used for designing a system:

1. Formulation of mathematical model.
2. Analysis of the sensitivity of a system with respect to its elements.
3. Analysis of the compatibility of various components and sub system.
4. Determination of stability of this system when subjected to various inputs.
5. Optimization of the system with respect to some preselected criterion.
6. Prediction of performance of the system.
7. Evaluation of system.

The energy system can be modeled through liner programming.

7.4.2 *Liner programming*

Liner Programming (LP) without a doubt is one of the most used methods of operation today. Problems from almost every phase of industry have been formulated successfully as Liner Programme Model (LPM). Application from oil and gas industry include the initial extraction of crude oil and gas from earth, the determination of the optional product mix when blending gas line and final phase when the refined product is delivered to the consumer. Other applications include the use of liner programming to optimize the mixing of livestock food, determining weaks menu in minimum cost while meeting all nutrients and take requirements, transportation of goods production scheduled in a manufacturing unit, hiring of steward, flight scheduled of airlines and allocation of TV time for advertisement assignment of job to people. The list goes on an on.

The linear programming model involve a liner function several variable to be maximized or minimized (be generally call it OPTIMIZATION) subject to a set of linear constraint and non negative restriction on variables.

Maximise (or Minimise)

$Z=C_1X_1+C_2X_2+ C_3X_3+............+ C_nX_n$

Subjected to -

$a_{11}X_1+ a_{12}X_2 \; +..............+a_{1k}X_k(*)b_1$

$a_{21}X_1+ a_{22}X_2 \; +..............+a_{2k}X_k(*)b_2$

$a_{31}X_1+ a_{32}X_3 \; +..............+a_{3k}X_k(*)b_3$

$a_{m1}X_1+ a_{m2}X_2 \; +..............+a_{mk}X_k(*)b_m$

All Xi > 0

Where 'Ci' is known as cost co-efficient of xi, 'xi' is an unknown variable a_{ij} is known constant 'bj' is known constant and (*) means,>, or fro each constant.

For Rural Energy Planning, the objective function is of cost minimization, or energy output maximization, subjected to fixed costs as well as variable costs of the total energy system.

Here is an example of formulation of a Liner Programming Model (LMP)

Case: For a village biogas center two type of biogas plant are available. The type first is of $6m^3$ and type second is on $4m^3$. Ten and eight bags of cement are required for construction of plant of type one and second respectively.

Fifteen and eight labours are required for construction of first and second plants, respectively. The total cement bags available are 82 and labour hours 107 only. How many plants of each type be constructed for achieving maximum gas output.

LP Model for the case -

Maximise $Z = 6X_1 + 4X_2$

Subject to $10X_1 + 5X_2 \leq 82$

$5X_1 + 8X_2 \geq 107$

$X_1X_2 > 0$

7.4.3 Solution for linear programming models

There are basically two approaches used for finding solution of Linear Programming Models.

(i) *Graphical Solution:* It's use is limited to a Linear Programming Model involving only two variables.

(ii) *Simplex Method:* The simplex technology is converging, iterative process. It start with a feasible solution and is so designed that it moves only in the direction of other feasible solutions which are better than original. If an optimum solution does not exit. The technique will cause a convergence to a solution in finite number of steps.

The solution of Linear Programming Model (LP Model) using simplex method can be found either manually or by using computers. While solving manually a large number of variable cannot be taken. Presently, computers are preferred in which large number of variables can be taken to make a model more prefects to solve LP models.

A number of commercial and special package are available in the market which can be used to solve LP models.

A computer package has been developed by Rural Energy Division of Planning Commission, Govt. of India. It consist of Macro Model for Rural Energy Sector which have a perspective plan for 15 year period from 1985 - 2000. it provide the optimum mix of different energy options: conventional, non-conventional, renewable and non-renewable etc. to meet the need for substance (cooking, lighting, heating) and production activities(agricultural and non-agricultural). The optimum supply mix would be determine by minimizing the cost for meeting the demand after taking into account various constraints on availability of resources, growth of technologies and regional constraints. The computer model also helps in micro level planning and provides the frameworks for integrated planning at block, district and State levels.

The optimization technique has been used for planning. The model is successfully tested for planning of energy for (i) cooking (ii) lighting (iii) heating and (iv) irrigation.

The objective function is of the cost minimization, which minimize the annualized fixed cost as well as the variable operation cost of the total energy system.

The model can take 200 variable and 264 constraints. The robustness of the model has been checked by carrying out sensitivity analysis. The Integrated Rural Energy Planning computer model has been found capable of analyzing the effect of policy measures on the optimum mix of different energy technology. The model uses the following unit;

Energy flow	: MWN per annum
Variable cost	: Rs. per MWN
Annualized fixed cost	: Rs. per annum for each supply technology

There are 264 constraints in the model which can be divided into six types as below:

1. Capacity constraints
2. Demand constraints
3. Supply constraints
4. Dynamic constraints
5. Fuel mix constraints
6. Technology constraints

8

Utilisation of Solar Energy for Thermal Application

8.1 Introduction

The concept of social justice led to a raise the standard of living of the masses. This is only possible by meeting the energy needs of industry, agriculture, transport and domestic use. The energy of demand has been increasing tremendously all over the world science mid - nineteenth century due to rise in industrial and agricultural activities and was being met to a great extent by fossil fuels.

Three factors led the mankind to look for alternative renewable sources of energy like solar, wind, hydro etc. (1) there is a danger that the rate at which the non renewable sources of energies are being consumed presently may lead to a global shortage in the near future. This has already happen in case of oil. (2) Use of fossil fuel involves emission of toxic gases polluting the environment causing health hazards of living being. and (3) the demand and supply gas of energy is continuously widening, thus necessitating to exploit renewable sources of energy.

Out of various renewable energy sources, solar energy has been a promising one and lot of development has been taken in its technology for various uses. The numbers of technology engaged in developing solar energy for meaningful use are aiming to achieve the following objectives-

1. Efficient conversion system to use the solar energy optimally for the required use.
2. Solar energy devices for various purposes should be economical.
3. Solar energy devise should be of low operation and maintenance cost.
4. Solar energy devices should not be inconveniently heavy to use.

Optimization of the above requisites can only lead to a successful solar energy system which can replace the traditional way of consuming energy. With the long term strategy in mind, the solar energy should be harnesses though useful devise for augmentation of thermal and electricity demand contributing largely to the development of the country.

Solar energy can be harnessed to useful propose by two routes (a) Thermal and (b) Photovoltaic.

A simple method of harnessing solar energy is by solar thermal conversion system that collects and utilizes the sun's energy in the form of heat. The heat or thermal energy may be used for several applications.

8.2 Solar Cooking

Cooking food with solar energy is a simple application of solar thermal technology. Solar cooker works on solar energy and need no fuel, emit no smoke, no soots spoil the cooking utensils as well as kitchen walls and it keeps the environment cleans.

The principle methods of cooking food are boiling, frying, roasting and baking. In conventional food habit the boiling technique is employed for preparation of pulse, vegetable and rice and during this process the temperature of food being cooked is about 100°C. For other methods of cooking higher temperatures are required. For frying and boiling, heat is supplied from all the sides of the material being cooked and heat to the food is being transferred through convection and radiation. In most of food, water is already present in raw material. Further, some water is also added to get homogenous mixture of food for cooking. In such cases, once the boiling temperature (close to 100°C) is reached, not much heat is required except the heating rate should be equal to the rate of thermal loss from the vessel. The thermal losses from the cooking vessels are: evaporation loss from the food, radiation and convection loss from the surface of the cooking vessel etc. Hence in the design of solar cooker attention be given to following points: -

1. It should provide sufficient heat for quick boiling of water so that food may be cooked as soon as possible.
2. Heat losses are minimized by incorporating insulation on the sides of vessel and keeping vessel covered with a lid.
3. The temperature requirement for cooking food by boiling is about 100°C, but to have high heat transfer rates, the temperature of the heat source should be high. Therefore a solar collection of sufficient size be adopted for supplying heat higher than 100°C or so.

There are three broad categories of solar cookers available for cooking of food. The classifications of solar cooker are given in Fig. 8.1.

8.3 Direct or Focussing Type

As the name indicates in this type of solar cooker same kind of solar energy concentrator (reflector) is used, which ultimately reflect sun's rays to a common point (focus or focal point) on which a cooking pot is placed. In this type of solar cooker, the cooker utilizes the direct radiation only.

Presently, there are two commercially available solar cookers in this category, which are as follows:

(a) Dish solar cooker

It is concentrating type parabolic dish solar cooker with aperture dimension of 1.4 meter and focal length of 0.28 meter. It is suitable for individual and also for small establishments like dhabas, teashops etc. The reflecting material used for fabrication of this cooker is uniodized aluminium sheet, which has a reflectivity of over 75%. The reflecting bowl is paraboloid dish made of reflecting sheets supported on suitable rings for holding them in fixed position. The sheets are joined in such a way that they automatically form the parabolic shape. The structure and frame of the bowl is strong enough so that the reflectors do not get deformed while turning it in various directions.

Second important part of this cooker is reflector stand, which is made of mild steel having powder coating for better durability. The stand is such that it can rotate reflector 360° around the horizontal axis passing through the focus and the center of gravity. It is also rotated around the vertical axis so it can adjust the cooker in the direction of sun. The tracking of the cooker is manual and has to be adjusted in 15 to 20 minutes during cooking time. The cooker has a delivering power of about 0.6 KW which can boil 2 to 3 liters of water in half an hour. The temperature achieved at the bottom of the vessel is around 350 to 400°C, which is sufficient for roasting, frying and boiling. This cooker is having a thermal efficiency of around 40% and can meet the needs of around 15 people and used from one hour after sunrise to one hour before sunset on clear days (Fig. 8.2).

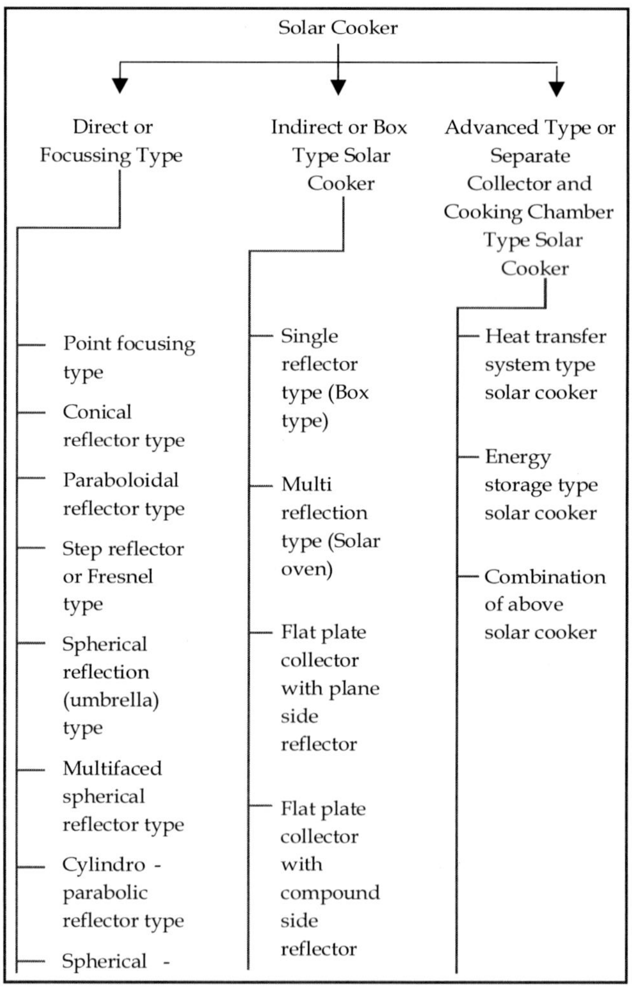

Fig. 8.1 Different Type of Solar Cooker

Fuel savings with dish solar cooker:

Instantaneously power available at the bottom of cooking pot :

$= I \times A_c \times \rho \times g$

where

I = Direct solar radiation falling on the cooker

A_c = Aperture area of the cooker

ρ = Reflectivity of the reflecting sheets

g = Intercept factor of sun rays at the cooking pot

In our case instantaneous power

= 600 watt / sq.m x 1.54 sq m x 0.75 x 0.85

= 0.6 KW

Assuming 300 clear days with 7 hours of average sunshine per day:

Total heat energy that could be available at the cooking pot during the year

= 0.6 kW x 300 days x 7 hrs/day

= 1200 kWh

Heat energy available from 1 LPG cylinder after lighting the gas

= W x C x h

= 14.2 kg x 10800 Kcal / kg x 0.60

where

W = Wt of cylinder,

C = Calorific value of gas

h = Burning efficiency of LPG stove.

Maximum LPG cylinders that could be saved on using the cooker continuously during available sunshine = 12 No.

(b) Paraboloid Type Community Solar Cooker for Indoor Cooking

It is possible to cook food within the kitchen with the help of this cooker. In this cooker a 7 sq.m large reflector standing outside the kitchen reflects the solar rays into the kitchen through an opening in its North wall while a secondary reflector further concentrates the rays on to the bottom of the pot painted black.

It is possible to cook the food of about 40 to 50 percent with the help of this cooker. The temperature attained is so high (400 °C) that the food could be cooked in a shorter time unlike box type solar cooker.

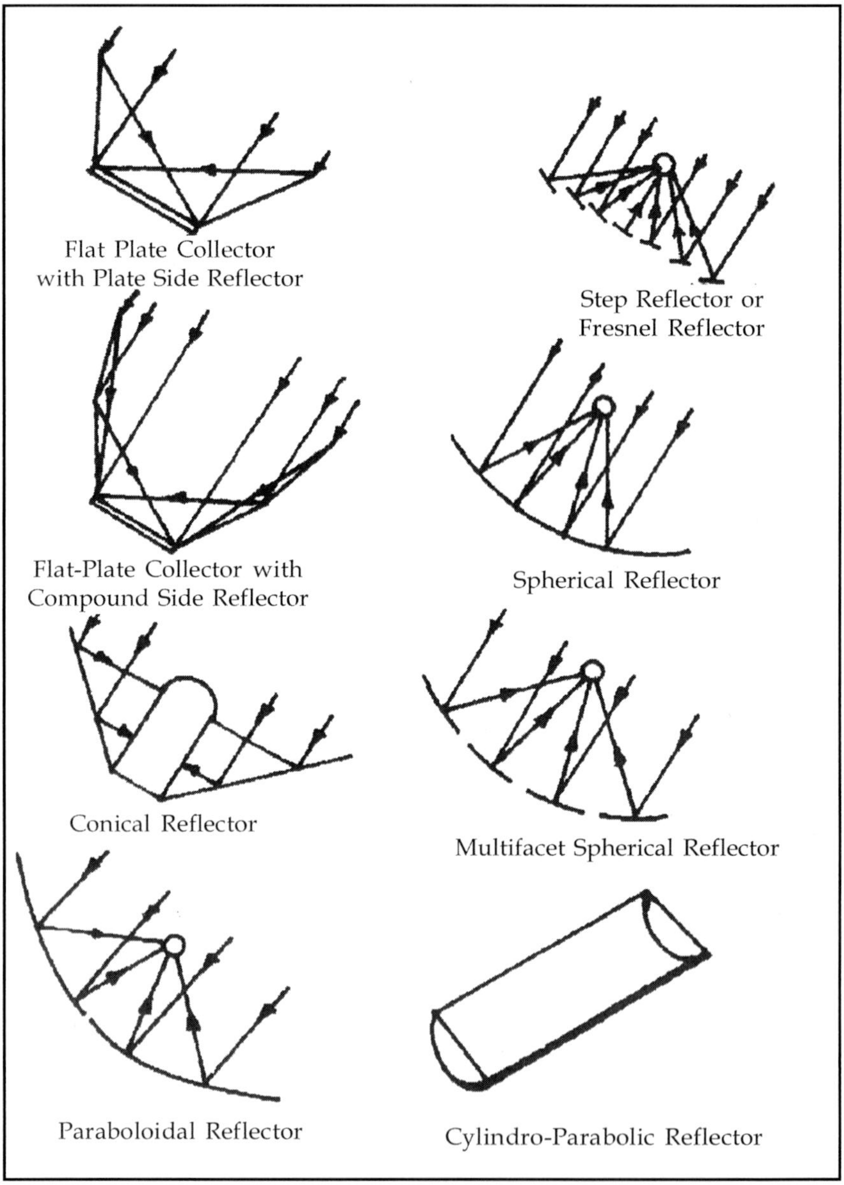

Fig. 8.2. Various type of Focussing Reflectors used in Direct type Solar Cooker

The major components of this cooker are as follows: Reflector frame, stand, rotating support, tracking device and secondary reflector.

The reflector frame, which is a primary reflector, is placed on a parabolic frame made out of metallic square tubes. It is a rigid structure of mild steel material. This reflector is made out of aluminium facet (plates) / rods on which reflecting mirrors of glass / acrylic fixed which reflects the light towards secondary reflector. The secondary reflector is essentially a reflecting sheet so curved in a shape that it reflects the incoming rays towards bottom of the vessel.

This type of solar cookers is also having a rotating support, which is meant to rotate the frame along with the sun so that the solar rays are always directed on the focal point within the kitchen. The movement of the rotating support on which the frame is fixed is done with the help of a tracking device which is pendulum clockwork made out of air blower and cycle parts. The cooker is to set once a day is the morning and thereafter for rest of the time the clockwork keeps on rotating the reflector automatically.

8.4 Indirect or Box Type Solar Cooker

Box type solar cooker is a typical example of solar energy devices, which demonstrate application of good reflector, absorber and transmitter.

In this type of solar cooker direct interception of solar radiation through glazing and indirect interception through reflection from plane mirror is made on the black body (insulated hot box) where raw material is placed for cooking. The insulated hot box may be square, rectangular or cylindrical in shape, which is painted black from inside with double-glazing for direct interception of solar radiation. The indirect entry of solar radiation through reflection is provided by single or multiple reflectors. Here the adjustment of cooker toward the sun is not so frequently required as in the case of direct type of solar cooker. It is slow cooker and takes long time for cooking and many of the dishes require roasting and baking can not be prepared with this cooker.

The commercially available box type solar cooker is illustrated in the Fig. 8.3. It has following important parts.

a. The outer box

The outer box of a solar cooker may be made of wood, iron sheet or fiber reinforced plastic having suitable dimensions, which accommodate black body (inner box) and insulating pads.

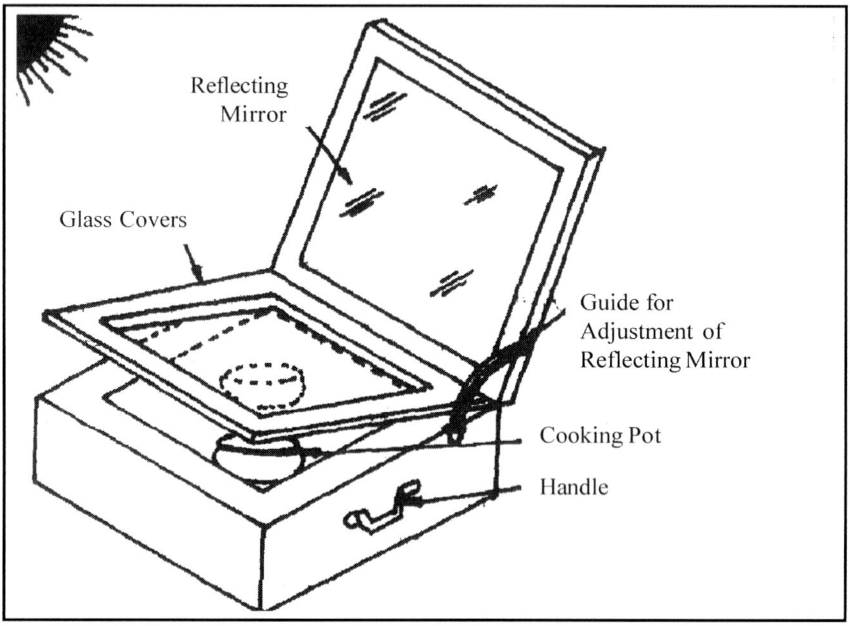

Fig. 8.3. Simple Box Type Solar Cooker

b. The inner box

The inner box may be made from G.I. sheet or aluminum sheet. All the four sides and the bottom of the inner box, which are exposed to the sun, are coated with black board paint for absorbing maximum amount of solar radiations. In fact, it is acting as black body, which absorbs energy directly intercepting through glazing and secondary reflecting form the plane mirror.

c. Insulation

The hot box must be thermally insulated so that heat gain through solar radiation be effectively used for cooking purpose. Therefore, the space between outer box and inner box must be filled with a quality insulating material such as glass wool, thermocole etc. A suitable thickness of good quality insulating materials maintained the temperature of inner box. Cavity between glazing act as a insulator, which prevent heat losses from the inside box to top of the surface.

e. Plane Mirror

A plane mirror is attached to the cooker, so that it enhances the entry of solar radiation by about 50 per cent in the inner box. In fact plane mirror acts as reflector and it increases the radiation input on the absorbing surface.

f. Cooking containers

These containers with covers are made of aluminum or stainless steel and having dull black paint on their outer surface so that maximum amount of radiation can directly absorbed.

8.5 Multi Reflector Type

Another box type solar cooker with a three-step reflector, consisting of plane mirrors is also available in India. The folding three-step asymmetric reflector box cooker retains the essential features of a box cooker and fitted with a three-step asymmetric reflector. As the reflector is adjusted relative to the box window, the distance of the mirror from the box window changes. Therefore this design will not give optimum performance in different seasons of the year. Optimum performance can be obtained if the mirrors also are hinged together so that their relative angles can also be changed. This arrangement is done in a two-step reflector box cooker. In this design the two reflecting mirrors are hinged with each other and the width of each mirror is about 1.25 times the width of transparent window. This folding two-step asymmetric system is designed to give a concentration of 1.89 in summer and 2.0 in winter.

A solar oven is similar to a hotbox cooker using an insulated box and multiple reflectors. The area of the box is kept as small as possible while the reflector area is large. The major difference is in the mechanism of directing solar radiation to the cooking area. In the hot box cooker direct solar radiation penetrates directly through the glass window while in a solar oven additional radiation after reflection also penetrates through the glass window. Due to enhanced solar radiation penetration and decreased cooker area for heat losses, the temperatures inside solar ovens can attain quite high temperature sufficient for cooking all varieties of foods. The mirrors are generally fixed at certain angles relative to the glass cover and the whole oven is oriented and tilted in the direction of the sun.

8.5.1 *Advanced type of solar cooker*

It is a separate collector and cooking chamber type solar cooker, where solar energy is collected at a separate place with the help of either a flat-plate or focussing collector and then this stored heat is transferred to the cooking vessel placed either at a separate place or indoor. Further, here the cooking in some cases can either be done with stored heat or the solar heat is directly transferred to the cooking vessel in the kitchen.

Now a day solar steam cooking system, which employs automatic tracking solar, dish concentrations are also available. These cookers can convert water into high-pressure steam. The steam thus generated is being used for cooking purposes. In India number of unique solar steam cookers have been installed which include solar steam cooker for preparation of 7500 persons at a time at Saibaba Temple Maharastra, steam solar cooker for 10,000 person at Mount Abu Rajasthan and cooker for 15,000 person at Trirupati Balaji Temple.

8.5.2 *Testing of solar cooker*

The box type solar cooker can be tested for its performance evaluation as per IS 13429 (Part 3), 1992. In this case figure of merit F_1 and F_2 are to be calculated, which should be greater than 0.12 and 0.25, respectively.

Figure of merit F_1 is calculated through thermal performance test under stagnation temperature condition, whereas figure of merit F_2 is calculated through thermal performance test under heat up condition.

The thermal performance test under stagnation temperature condition should be started in the following steps in the morning before 10.00 hr so that the stagnation temperature is achieved near solar noon condition. Firstly place the solar cooker in open sun condition. Remove the cover reflector of the solar cooker. Monitor the temperature of inside solar cooker using a RTD at an interval of 5 minutes continuously. When the cooker temperature has reached at steady state, note down the final cooker inside temperature (T_{pa}) and the corresponding outside ambient air temperature (T_{sa}).

The steady state conditions shall be defined as a 10-minute period when cooker temperature is constant within ± 1°C. Solar radiation is constant within ± 20 W/m^2. Ambient temperature is constant within ± 0.2 °C. and Solar radiation is greater than 600 W/m^2.

Calculate F_1, which is defined as the ratio of optical efficiency to the heat loss coefficient as follows:

$$F_1 = \frac{\alpha\tau}{U_L} = \frac{(T_{pa} - T_{sa})}{G_s}$$

where

τ = Transmission of glass

α = Absorvitivity of cooking tray

U_L = Heat loss coefficient of the cooker

G_s = Solar radiation during steady state

T_{pa} = Final Cooker inside temperature

T_{sa} = Outside ambient air temperature corresponding to stagnation temperature (T_{pa})

Thermal performance test under heat up condition is carried out as follows:

Weigh the empty cooking pots and then fill with water at 50 ± 5 °C. Reweigh and calculate the mass of water, place the pots in the cooker from which the mirror has been removed.

Place a temperature probe of data logger in the largest of the cooking pots with the measuring tip submerged in the water. The ambient temperature and wind speeds are measured through out the test. The test started in the morning between 10.00 h and 11:30 h of local solar time. The radiation and temperature are measured by spot checks, at an interval of five minutes. Constant monitoring at 30-second intervals is maintained with averages of radiation recorded over 2 minute's intervals. Following measurements are taken:

(i) Water temperature is measured along with the exact time of that measurement recorded.

(ii) The data recording should be continued until the water temperature exceeds 95°C.

(iii) The initial and final temperature / time data pairs should be chosen. The initial temperature between 60C to 65°C and the final temperature in between 90°C to 95°C is selected. These are noted as T_{w1} and T_{w2} respectively and the corresponding times t_1 and t_2 respectively.

(iv) The average air temperature (T_a) between the time t_1 and t_2 be calculated.

(v) The radiation recorded between the two points during observation should not vary by more than ± 10 W/m^2 and always exceeded 600 W/m^2.

(vi) The average radiation (G) over the time t_1 and t_2 is calculated.

(vii) The second figure of merit F_2 was calculated as follow:

$$F_1 = \frac{F_1.(MC)_w}{A(t_2 - t_1)}.\log_e \left[\frac{1 - \frac{(T_{w1} - T_a)}{F_1.G}}{1 - \frac{(T_{w2} - T_a)}{F_1.G}} \right]$$

where

F_1 = First figure of merit from stagnation test

$(MC)_w$ = Product of the mass of water and specific heat in J/°C

A = Aperture area of the cooker of cover plate in m^2

$(t_1\text{-}t_2)$ = Time taken for heating from T_{w1} and T_{w2} in seconds

T_a = Average air temperature over time period $(t_1\text{-}t_2)$ in °C

G = Average radiation over time period $(t_1\text{-}t_2)$ in W/m^2.

8.6 Solar Water Heater

Water heating is one of the simplest applications of solar energy. Solar water heater offers a number of advantages such as:

1. Simple in construction and installation.
2. Almost no maintenance and operating cost.
3. It saves time and also energy for heating water.
4. Easily retrofittable to existing houses as per requirement.
5. Economical viable as compared to electric water heater.
6. Moderate temperature required, domestic affairs could easily achieved.

There are two basic types of solar water heaters: -

1. Collector coupled to storage tank.
2. Collector-cum-storage system.

First type of solar water system can extract heat from solar collectors in two ways, namely through natural convection (also known as thermosyphon effect) and through a forced flow of water using an electrically operated pump. Collector coupled to storage tank type of water heater is used, as domestic solar water heater where the maximum temperature required is no more than 70°C. A typical system consists of solar collector (front glazing), metallic absorber, back insulation and collector box, insulated storage tank (with or without heat exchanger), piping, controls and pump.

Here heating is accomplished by collection of solar radiation with flat plate collector mounted on south facing roof or walls. The collector is usually placed below the storage tank, cold water from the storage tank flows down to the inlet of the collector, gets heated in the collector and rises to the tank by thermosyphonic effect. A density difference created by the temperature gradient causes the fluid to flow up in the collector by the thermosyphonic effect (Fig. 8.4).

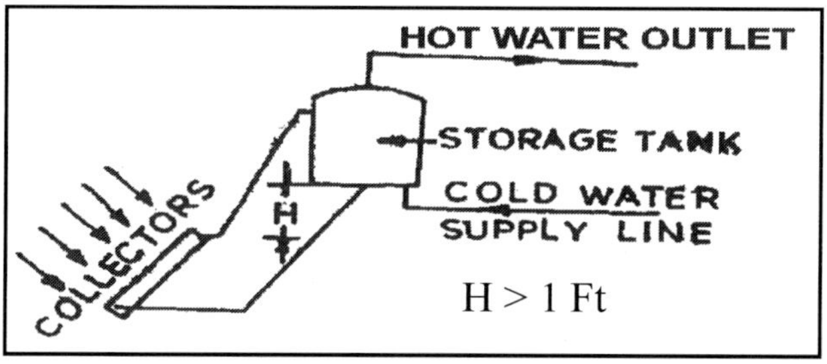

Fig. 8.4 Thermosyphon type Solar Water Heater (Direct Natural Circulation type)

The collectors are usually oriented south with an inclination angle equal to latitude of the place. For round the year use, the inclination of the collectors from the horizontal should be equivalent to the latitude. However for winter use, it is preferable to keep angle of inclination at latitude + 15°C. The required collector area for 100-liter hot water demand at 50°C is about 2 m^2.

In the cold climates, it is essential to mix anti-freeze materials in the heat extracting fluid circulating through the collector channels. In this case a heat exchanger may also be used and which prevent antifreeze fluid to flow directly through the collectors. It is known as indirect natural circulation type solar water heater (Fig.8.5).

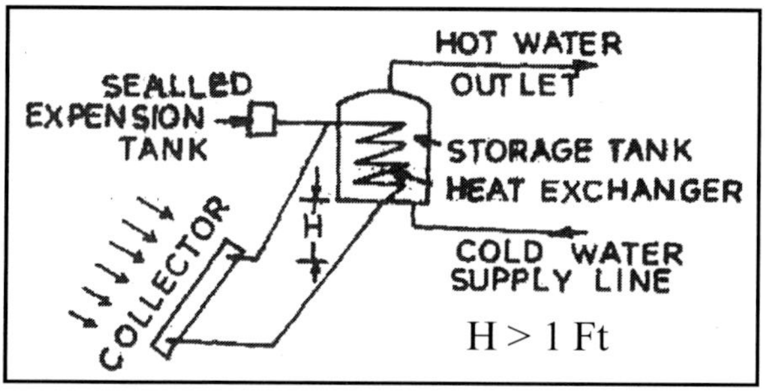

Fig. 8.5 Indirect Natural Circulation Solar Water Heater

In the case of collector-cum-storage type solar water heater, all the three functions i.e. solar energy collection, storage and control are combined into a single unit. Essentially, a typical collector-cum-storage water heating system consists of three main components - front glazing (solar flat plate collector), absorber sheet and insulated storage tank. Sun's ray passes through the front glazing and gets absorbed by the absorber, which further heats the water. The storage tank acts as a collector cum storage for hot water. A schematic diagram of system is shown in Fig. 8.6.

The utility profile of hot water by a user is an important factor to decide about the viability of solar water heater. This could be evaluated by using kilowatt meter, flow meter or energy meter for fuel consumption. If the load is continuous and when peak solar energy is available then solar water heater could be a good proposition. But to find out optimum size of the solar water heater, the load requirements on daily and on seasonal basis need to be known. This would help in deciding the collector area and also the storage capacity. This again would be limited by the open sunny space available for mounting of the collector field.

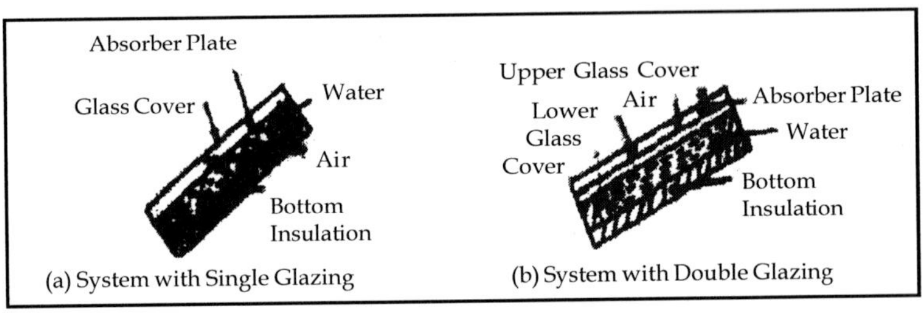

Fig. 8.6 Schematic of Collector-Cum-Storage

Normally, the solar water heaters are either installed on the ground or on the rooftop. For system to be installed on the roof top, roof construction and load bearing capacity, orientation of building accessibility, avoiding shadow due to nearby structures, plants/ accessories / tall trees, and availability of cold water must be ensured. Similar care is taken for the systems to be installed on the ground.

It is also possible to integrate the solar water heaters with the building design in the new construction. The solar collectors could either be made a part of the south-facing wall of the building or could be installed on the window/door overhangs, etc. Thus the hot water requirement in the building could be supplemented even without affecting the aesthetics of the building. Several such systems are in operation in the country.

Normally, four types of collectors are being used in solar water heaters for different temperature requirements. These are:

1. Unglazed polymer collectors for heating of swimming pool water.
2. Metallic collector with black painted absorber (e.g. copper tube/ copper sheet and copper tube aluminum sheet etc.) for temperature applications upto 60°C.
3. Collector with selectively coated metallic absorber for temperature upto 80-85°C and
4. Evacuated tubular collector for temperature above 80°C.

Further, the collector has to be chosen on the basis of the requirement and temperature of the hot water.

In the thermosyphon type of solar water heater, the flow of water from bottom of the tank through the collector of the top of the tank takes place because of the temperature gradient created in the collector under the sunlight. This configuration is being used in almost all the low capacity solar water heaters including those for domestic applications. Solar water heaters of capacity upto 5000 lpd have been installed in the country based on this principle and these have been successfully operating. Since there is no moving part in such a system, the chances of failure are very rare.

In the forced flow type of configuration, pumps with controls are added at appropriate place in the system to ensure the flow of water through all the collectors. This type of system can either be a single pass system where the water enters the collector array at one end and leaves at the other or a recirculation type system. Where water is taken from the storage tank circulated through the collector array and returns back to the tank repeatedly. Since in a single pass system the collector operates at comparatively lower temperature than in recirculation type. The system efficiency is better in the previous case.

Collector interconnection must be able to keep up expansion and contraction between the collectors, prevent leakage at collector couplings, and prevent galvanic corrosion, compatible with heat transfer fluid (water) and resistant to weather and outside exposure. Brass flanges with neoprine gaskets are normally being used.

Normally B class GI pipe is being used for piping. Wherever this type is connected to copper header, a corrosion protector (gasket) is to be used to prevent galvanic corrosion. The insulation on the pipe must be protected from water absorption and UV deterioration. Sealants are to be used over the insulation layer. The insulated pipe is to be supported properly.

The capacity of solar water heating system can be boosted by increasing collector area. For large capacity solar water heaters, the collectors are to be connected in either parallel-series arrangement or in series-parallel arrangement. The advantages in parallel-series arrangement are that the pipe length in series is reduced and also the thermal capacity of the system is reduced. This arrangement, however, has to have venting provision during initial filling and also for release of dissolved air. Whereas in the series-parallel arrangement, the piping length is the least and so is the thermal capacity but it must be ensured in this case that all the collectors running full with water, venting is effective and the flow rates are adequate to avoid over heating and blockage. (Fig 8.7).

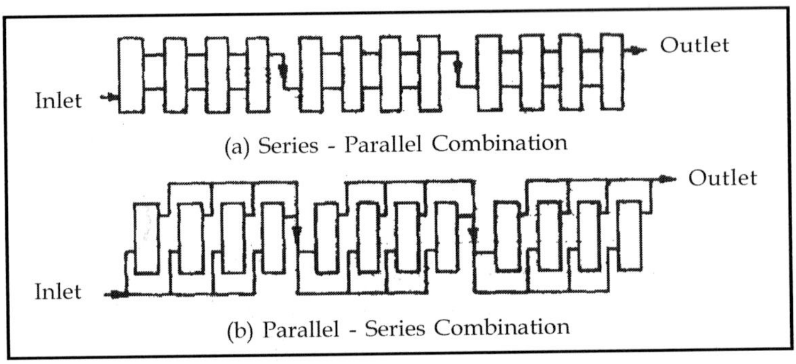

Fig. 8.7 Combination Techniques of Solar Collector

Heat exchanger must be used in case where the quality of water is such that the collectors are subjected to scale formation and also where freezing may occur during night. Counter flow heat exchangers and coil in storage tank type heat exchangers are normally being used in solar water heater.

For areas where night temperature drop below the freezing point, double loop systems with antifreeze mixture in the primary loop are to be used. The composition of the anti-freeze mixture depends on the minimum ambient temperature of the site. One can also use single loop design with drain-off facility. In such a design, the collector field is drained off completely when the minimum temperature is likely to fall below the freezing point.

Solar water heaters normally require minimum maintenance. This is particularly true for systems based on the thermosyphon principle. However, cleaning of the collector glazing once in a week is desirable for better performance of the system. In areas where scale formation is a problem, de-scaling need to be done once in a year. In case of black painted collector, repainting may be required after a period of two to three years, whereas this may not be required in case of selectively coated collector. In case of forced circulation system, the conventional components of the system such as the thermal sensors, relays and pumps might need regular check up and maintenance. The average cost of domestic solar water heater of 100-lpd capacity in the country is about Rs.15,000. Assuming 5 kWh as daily average solar insolation and 300 clear sunny days in a year, a domestic solar water heater can supply upto 1800 kWh of thermal energy per annum. Similarly, the cost of commercial and industrial solar water heater varies from Rs. 70 to Rs. 120 per liter based on the site conditions, capacity and temperature requirements. The saving again would be 400 kWh of thermal energy per meter square of the installed collector area per annum.

8.7 Solar Desalination

There is an important need for clean, pure drinking water in many developing countries. Often water sources are brackish (i.e. contain dissolved salts) and/or contain harmful bacteria and therefore cannot be used for drinking. In addition, there are many coastal locations where seawater is abundant but potable water is not available. Pure water is also useful for batteries and in hospitals or schools.

Distillation is one of many processes that can be used for water purification. This requires an energy input as heat, solar radiation can be the source of energy. In this process, water is evaporated, thus separating water vapour from dissolved matter, which is condensed as pure water.

Solar water distillation is a solar technology with a very long history and installations were built over 2000 years ago, although to produce salt rather than drinking water. Documented use of solar stills began in the sixteenth century. An early large-scale solar still was built in 1872 to supply a mining community in Chile with drinking water.

Mass production occurred for the first time during the Second World War when 200,000 inflatable plastic stills were made to be kept in life-crafts for the US Navy.

There are a number of other approaches to water purification and desalination, such as photovoltaic powered reverse-osmosis, for which small-scale commercially available equipment is available. These are not considered here.

In addition, if treatment of polluted water is required rather than desalination, slow sand filtration is a good option.

8.7.1 Energy requirements for water distillation

The energy required to evaporate water is the latent heat of vaporisation of water. This has a value of 2260 kilojoules per kilogram (kJ/kg). This means that to produce 1 litre (i.e. 1kg since the density of water is 1kg/litre) of pure water by distilling brackish water requires a heat input of 2260kJ. This does not allow for the efficiency of the heating method, which will be less than 100 per cent, or for any recovery of latent heat that is rejected when the water vapour is condensed.

It should be noted that, although 2260kJ/kg is required to evaporate water, to pump a kg of water through 20m head requires only 0.2kJ/kg. Distillation is therefore normally considered only where there is no local source of fresh water that can be easily pumped or lifted.

8.7.2 How a simple solar still operates

Fig 8.8 shows a single-basin still. The main features of operation are the same for all solar stills. The incident solar radiation is transmitted through the glass cover and is absorbed as heat by a black surface in contact with the water to be distilled. The water is thus heated and gives off water vapour. The vapour condenses on the glass cover, which is at a lower temperature because it is in contact with the ambient air, and runs down into a gutter from where it is fed to a storage tank.

8.7.3 Design objectives for an efficient solar still

For high efficiency the solar still should maintain:

- A high feed (undistilled) water temperature
- A large temperature difference between feed water and condensing surface
- Low vapour leakage.

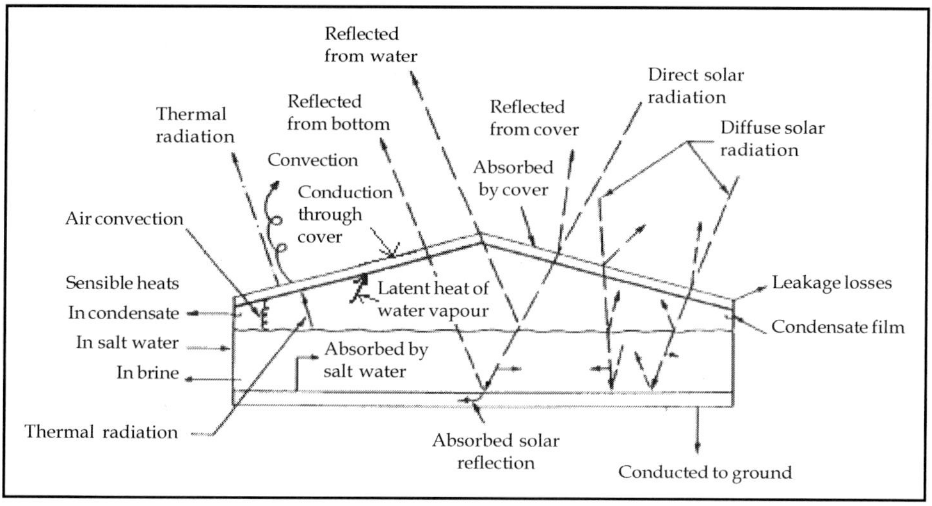

Fig. 8.8. Solar Still

A high feed water temperature can be achieved if:

- A high proportion of incoming radiation is absorbed by the feed water as heat. Hence low absorption glazing and a good radiation absorbing surface are required.
- Heat losses from the floor and walls are kept low.
- The water is shallow so there is not so much to heat.

A large temperature difference can be achieved if:

- The condensing surface absorbs little or none of the incoming radiation.
- Condensing water dissipates heat which must be removed rapidly from the condensing surface by, for example, a second flow of water or air, or by condensing at night.

8.7.4 Design types and their performance

Single-basin stills have been much studied and their behaviour is well understood. Efficiencies of 25 per cent are typical. Daily output as a function of solar irradiation is greatest in the early evening when the feed water is still hot but when outside temperatures are falling.

Material selection is very important. The cover can be either glass or plastic. Glass is considered to be best for most long-term applications, whereas a plastic (such as polyethylene) can be used for short-term use.

Sand concrete or waterproofed concrete are considered best for the basin of a long-life still if it is to be manufactured on-site, but for factory-manufactured stills, pre-fabricated ferro-concrete is a suitable material.

Multiple-effect basin stills have two or more compartments. The condensing surface of the lower compartment is the floor of the upper compartment. The heat given off by the condensing vapour provides energy to vaporize the feed water above. Efficiency is therefore greater than for a single-basin still typically being 35 per cent or more but the cost and complexity are correspondingly higher.

Wick stills : In a wick still, the feed water flows slowly through a porous, radiation-absorbing pad (the wick). Two advantages are claimed over basin stills. First, the wick can be tilted so that the feed water presents a better angle to the sun (reducing reflection and presenting a large effective area). Second, less feed water is in the still at any time and so the water is heated more quickly and to a higher temperature.

Simple wick stills are more efficient than basin stills and some designs are claimed to cost less than a basin still of the same output.

Emergency still : To provide emergency drinking water on land, a very simple still can be made. It makes use of the moisture in the earth. All that is required is a plastic cover, a bowl or bucket and a pebble.

8.7.5 Hybrid designs

There are a number of ways in which solar stills can usefully be combined with another function of technology. Three examples are given:

- *Rainwater collection.* By adding an external gutter, the still cover can be used for rainwater collection to supplement the solar still output.
- *Greenhouse-solar still.* The roof of a greenhouse can be used as the cover of a still.
- *Supplementary heating.* Waste heat from an engine or the condenser of a refrigerator can be used as an additional energy input.

8.7.6 Output of a solar still

An approximate method of estimating the output of a solar still is given by:

$$Q = \frac{E \times G \times A}{2.3}$$

where

Q = Daily output of distilled water (litres/day)

E = Overall efficiency

G = Daily global solar irradiation (MJ/m^2)

A = Aperture area of the still ie, the plan areas for a simple basin still (m^2)

In a typical country the average, daily, global solar irradiation is typically 18.0 MJ/m^2 (5 kWh/m^2). A simple basin still operates at an overall efficiency of about 30%. Hence the output per square metre of area is:

$$\text{daily output} = \frac{0.30 \times 18.0 \times 1}{2.3}$$

= 2.3 litres (per square metre)

The yearly output of a solar still is often therefore referred to as approximately one cubic metre per square metre.

The cost of pure water produced depends on:

- The cost of making the still
- The cost of the land
- The life of the still
- Operating costs
- Cost of the feed water
- The discount rate adopted
- The amount of water produced.

8.8 Solar Greenhouse

A greenhouse is quasi-permanent structure, covered with a transparent or translucent material, ranging from simple home made designs to sophisticated pre-fabricated structures, wherein the environment could be modified suitable for propagation or growing of plants. Materials used to construct a greenhouse frame may be wood, bamboo and steel or even aluminum. Covering can be glass or various rigid or flexible plastic materials. Depending on the covering material, different terminology have been used in the context of greenhouse structures as mentioned below:

Glasshouse: A greenhouse with glass as the covering material is called glasshouse

Polyhouse: It is a greenhouse with polyethylene as the covering material.

8.8.1 Plant environment and greenhouse climate

A plant grows best when exposed to an environment that is optimal for that particular plant species. The aerial environment for the plant growth can be specified by the following four factors:

1. Heat or temperature
2. Light
3. Relative humidity
4. Carbon dioxide

While plants have precise optimum environmental conditions for best growth, most are tolerant to variations in these conditions within some limits. However, permanent damage would occur when they are exposed to conditions outside these limits. At the same time plants are subjected to attack by pest and disease.

Greenhouse crop production protected against adverse environmental conditions and allows pest and disease to excluded or controlled. Besides providing portative enclosure. A greenhouse also acts as a 'Heat trap'. It admits solar radiation and converts this energy into heat by raising the temperature of the greenhouse air. While this is the basis of the greenhouse's ability to perform its tasks, it also affects others environmental factors.

Environmental conditions inside the greenhouse can be modified suiting to the potential growth of the plants. The extent of climate modification will, however, depend on the design of greenhouse and is generally related with it cost. Higher the capacity of greenhouse to modify its climate, higher is the cost of its construction.

The way in which a greenhouse gets heated exposed to sunlight is similar to heating of the earth's surface and its adjacent atmosphere. When solar radiation reaches the earth small portion is reflected back into the space while the remainder is absorbed at the surface raising its temperature. In the same way, when solar radiation reaches to greenhouse cover- surface, a small amount (normally - 15-20 per cent) is reflected back from the surface while the remainder transmitted to the interior. Plants, soil and other objects absorb most of this transmitted radiation and remainder is reflected as shown in Fig 8.9.

The absorbed radiation raise the temperature of absorbing surfaces and objects with the heat energy being immediately transferred to the greenhouse air by convection and evaporation, thereby increasing the temperature and humidity.

Greenhouse loose heat in three ways

1. Warm air moves out the greenhouse and is replaced by cooler outside air.
2. Heat is transferred through the covering material itself.
3. Heat is lost through the soil/floor

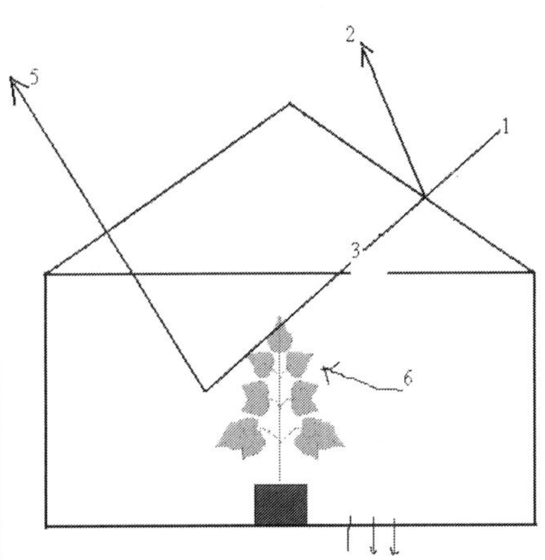

1 Solar radiation reaching the greenhouse

2 Solar radiation reflected from greenhouse surface.

3 Solar radiation transmitted to interior of greenhouse

4 Absorbed radiant heat energy conducted into surface.

5 Solar radiation reflected from internal surfaces.

6 Absorbed radiant heat energy transferred to the air by convection and conduction and evaporation.

Fig. 8.9. Solar Radiation Heats the Greenhouse

Increasing the temperature of the growing environment in a greenhouse is unavoidable, but it is also the most important function of a greenhouse. Enhanced temperatures accelerate plant growth, and allow sustaining plant growth even when outside ambient temperatures are unfavorably low. However, during summers inside temperatures rise higher than the optimum level and, therefore, cooling/ ventilation provision are necessary.

Most plant grows better within 60-80 per cent relative humidity (RH) of air. Low RH increases the evaporative demand on the plant, while high RH can depress this demand inhibiting the uptake of nutrients, particularly calcium. In general, the RH inside the greenhouse is higher than the outside, mainly due to transpiration load. Effective ventilation is required to control higher RH level.

In most part of the country, solar radiation is not a limiting factor for the plant growth. Light control inside the greenhouse can be affected conveniently either by shading or by supplementary lighting whenever required. Grower in northern India should, however be careful in monitoring light level in winters especially during prolonged foggy conditions. In peri urban areas, particulate pollutants get deposited on the plastic roof thereby reducing the light transmission significantly. This problem is compounded in winter. In such conditions, it is necessary to wash the roof frequently to maintained adequate light level inside the greenhouse.

Plants use carbon dioxide from the atmosphere for photosynthesis. Carbon dioxide concentration inside the greenhouse in the early morning is always higher the outside. With the onset of sun, this level quickly depletes and goes down the normal level during the day if adequate air exchanges are not maintained. Carbon dioxide enrichment is generally accomplished by burning suitable fuel like propone.

8.9 Solar Drying

Drying is a universal method for preserving the food and thereby helps in storage and easy transportation, because food becomes lighter due to moisture removal. Traditionally food products are dried by spreading in open sun in thin layer, this method is known as natural or open sun drying. It is economical and simple method of drying, however it suffers following drawbacks:

1. There is no control on the rate of drying. Product may be over dried or under dried during open sun drying.

2. There is no uniform drying. It results poor quality of products.
3. There are chances of deterioration of food in both slow and fast drying.
4. There is no protection of products against rain and dust storm etc.
5. There may be considerable damage due to birds, rodents insects and pests in open sun drying.
6. There may be chances of loss of germination power and nutritional changes in the seeds.

Therefore, in order to have product quality wise and quantity wise better and to prolong storage capabilities and improver transportability, controlled mechanism of solar drying system is required. Essentially it has provision of boosting temperature / heat over the product to be dried and has sufficient air flow rate for taking away moisture from the product or system to outside in the atmosphere. Solar drying is one of the direct use of solar heat, which involves removal of moisture from the substance through the application of solar energy. The removal of moisture requires only low temperature heating, which can be met easily by absorbing solar energy by the surface. This increases the temperature of the air inside the dryer. The moisture produced from the drying product is usually carried out along with the exhaust air. The characteristic of solar energy is good for the drying at low temperature i.e. high flow rates with low temperature rise. The intermittent nature of solar radiation can not effect the drying performance at low temperature. Further, solar energy is available at the site of use and saves transportation cost. Therefore, viewing to simplicity of drying through solar energy, various type of solar dryer has been designed and developed (Fig. 8.10).

Natural convection type solar dryer essentially consists of an enclosure for keeping the products to be dry with a transparent cover placed over the enclosure. The internal surfaces of the enclosure are painted black. The evaporation of moisture from the product takes place due to direct absorption of solar radiation by the product as well as transfer of heat by the internal surface, which get heated by the radiation incident on them. Here, ultimately removal of moisture takes place naturally created draft. In the forced circulation type solar dryer, air is heated through collectors and is forced on to the drying material. Such dryers are comparatively efficient, faster and can be used for drying large agricultural products.

Dryers are also classified as direct as well as indirect type of solar dryer. In the direct type solar dryer, the material to be dry is put on trays in the dryer and direct entry of sun through a suitable transparent cover is made and finally moisture after getting evaporated goes out. Where as in the indirect solar dryer some air heaters are used through which air gets heated and passes over to the drying material. Similarly direct cum indirect type of solar dryer, directly sun radiations are allowed inside the tray and also heated air from outside air collectors is used for boosting output of the dryer. There are varieties of each of these dryers. Few solar dryers are shown schematically in Fig. 8.11.

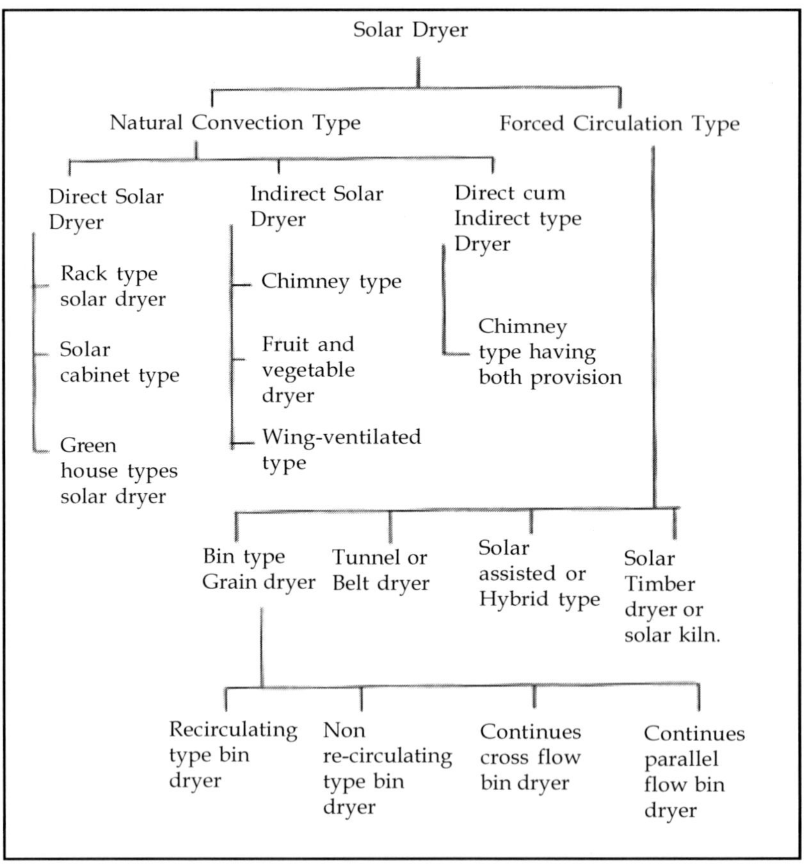

Fig. 8.10. Different type of Solar Dryer

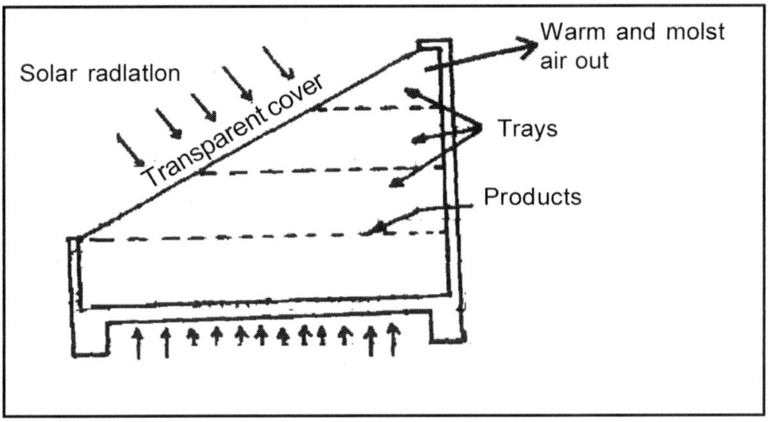

(a) Direct Natural Convection Solar Dryer

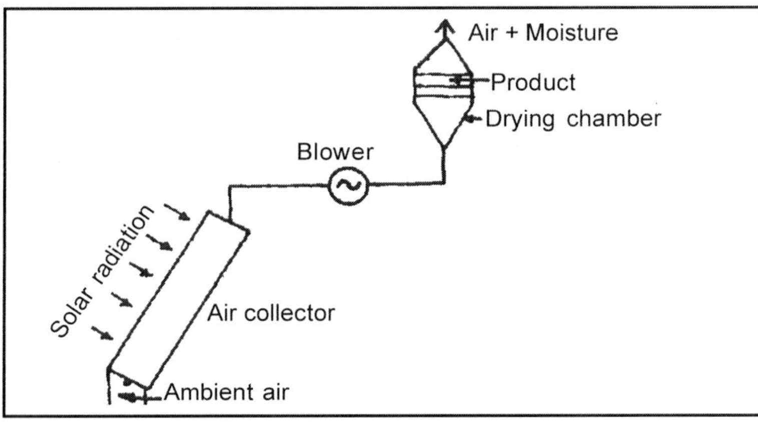

(b) Indirect Forced Convection Solar Dryer

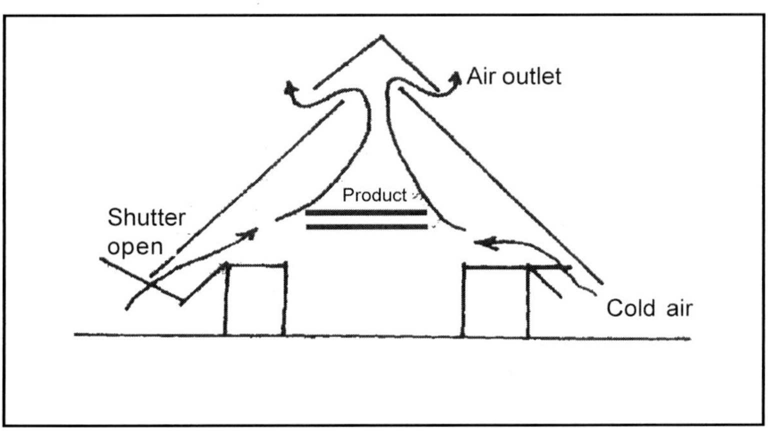

(c) Greenhouse Solar Dryer

Fig. 8.11 a-c. Various type of Solar Dryer

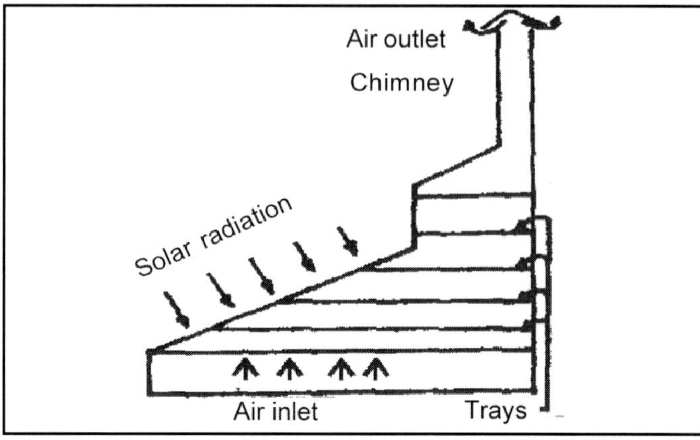

(d) Natural Convection Direct cum Indirect Solar Dryers

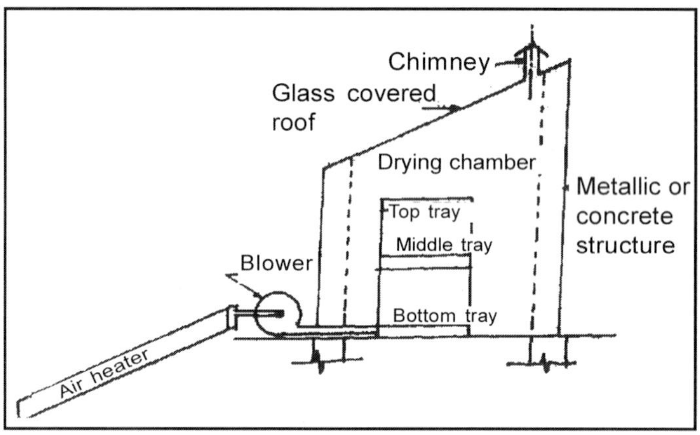

(e) Forced Convection Direct cum Indirect Solar Dryer

Fig. 8.11 d-e. Various type of Solar Dryer

8.10 Solar Space Heating

Basically for heating or cooling the space using solar energy, the solar energy is to be collected, stored and distributed properly in the space to be heated or removed from the space to be cooled. There are two basic approaches for solar houses, there are:

(a) Active solar house technology

(b) Passive solar house technology

In active approach, the solar energy is collected and stored in some separate solar energy collector and then energy is distributed in the space where heating is to be done using electrically operated pumps and fans coupled with radiator etc. The storage of solar energy in active solar house may be in the form of storage as sensible heat or storage as latent heat or chemical heat storage. As per requirement of heating the size of solar collector and storage system be decided. Generally the active concepts of solar houses are used for heating purpose. Active concept can also be used for space cooling applications. Solar vapor compression and absorption system can be used for this purpose, essentially, this includes incorporation of solar heat collectors, thermal storage system, auxiliary heat supply system (to compensate intermittent nature of sun energy) and control systems. Further, there are three ways of solar space heating i.e. solar air systems, solar liquid systems and solar heat pump systems.

The passive solar house technology may be used for space heating as well as for cooling. In this approach, orientation and layout of the buildings is so arranged, that three function of solar houses i.e. solar energy collections, storage and distributions is done by natural means and no electrical, mechanical or electronics controls are used. In the passive house, all the building components such as walls, roofs, windows, partitions, ventilation's walls etc. are so selected and arranged architecturally, so that solar collection, storage and distribution is managed automatically. In this approach, solar energy oriented parameters such as solar radiation's, outside air and internal metabolism, air flow rate are used for heating purposes, where as sky & space temperature, out side wind and wet surface are used for cooling purposes. The building materials (stone, bricks, cement etc.) and the various thermal processes (thermal radiation, natural and forced convection, conduction, air stratification, evaporation, thermosyphoning etc.) are so integrated along with building components and thereby efficient & convenient atmosphere is attended with the house. Depending upon site, climate & activity and particular needs, a proper balance is needed among building components, building elements and thermal processes. The passive heating & cooling process follows a sequence of several phases. Although the overall process might be described in different ways, the major steps are:

1. Identifying suitable building components for desired cooling and heating.
2. Identifying and establishing the criterion of acceptance of different building elements (stone, bricks etc.)
3. Incorporating suitable thermal process
4. Integrating optimally step (1), (2) & (3)

8.10.1 *Active solar house technology*

In the active solar houses, separate solar collectors are needed for heating the space as per requirement. The heating of house in winter is one of the major requirements for all types of houses. The technique of heating house with the help of some gadgets based on solar energy is an ancient technology. Basically, the active solar houses consisted of following major parts (Fig. 8.12).

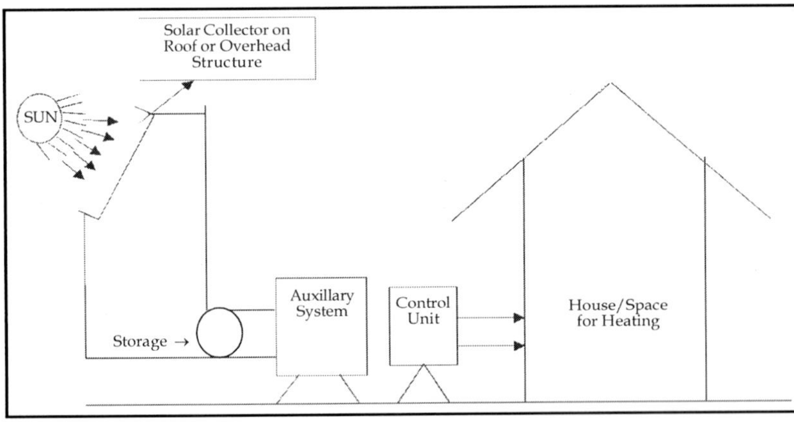

Fig.8.12 Simple Configuration of Active Solar Houses

(i) Solar Collectors

Some types of solar collectors are used for heating a fluid. These solar collectors may be air collectors, liquid collectors or heat pump type. In the air collectors, the air is heated and after words the heat is transferred to living space from storage unit by means of air or liquid whereas in liquid collectors, any suitable liquid preferably water is used for heating purpose.

(ii) Storage System

The collected solar energy in turn is stored in suitable structure either based on sensible heat storage concept or latent heat storage concept. The chemical storage systems are also employed depending upon its availability and cost involvement. In fact, storage devices are used to store heat for use at night and on intermittent days.

(iii) Auxiliary Heating System

Viewing to intermittent nature of sunlight, it is always advisable to have auxiliary system for night or for seasons when solar energy is not available. Basically auxiliary-heating system is provided to supply heat when required or to compensate performance of solar heating devices. However, such system generally turns out to be costly.

(iv) Controls

Some sort of controls either electrical or mechanical or electronically operated are used for supplying stored energy into space for heating purpose.

Basically, solar houses are meant to get more solar energy in winter and less in summer, so that this reduces the heating and cooling load for the building. Therefore, active solar houses, apart from its usual components, must have following pre-requisition:

It must be adequately insulated, so that optimum resistance to flow of heat is maintained.

- It must be properly oriented to take advantage of nature's gifts.
- It must be optimally glazed, so that heat acquition is maintained naturally as well as heat radiation should not go through openings.
- It should be adequately sealed against air leakages, so that there would be minimum heat losses.
- It must be made up of appropriate materials and combinations of materials, which should have adequate properties of absorption, admittance, storage, etc.

The addition of above mentioned component in the active houses boosted the overall performance of the system and thus reduces cost of cooling load in summer. Therefore, in the designing of active houses, proper planning is made in construction of house apart from its general parts.

Apart from above essential components of active solar heating systems, there are many other additional components that are required as per climatic and operating conditions, these are:

- A heat exchanger is required in the collector and storage unit, this required when the fluid in collector unit and storage unit is different.
- Drain-down type of collector, essentially for liquid/water collector in the places where chances of water freezing in there in the night.

- Multi-layered and multiplex storage system, storing the heat at different temperature for different applications and for different periods.
- A device for exhausting the surplus heat for preventing damage against boiling of water.

8.10.2 Types of active solar houses

Basically there are three type of active solar house, which are as follows:

- Active solar house equipped with air collector.
- Active solar house equipped with liquid collector.
- Active solar house with heat pumps mechanism.

All these three technologies are widely tried and are used for space heating. However, every system has got its own advantage and disadvantage, therefore, depending upon its technical feasibility, economical viability and efficiency, one has to adopt a appropriate type of system. The right choice of a active system depends in general, on the following factors:-

- The expected time of heating and expected time of solar energy availability on the place.
- The geographical locations of the place.
- The nature of the load to be connected i.e. complete house heating, partial house heating and single room heating.
- The degree of reliability needed for the process.
- Availability of scope of auxiliary space heater & the manner in which it would supplied heat.
- An economic analysis that determines how much of the total usually, annual loads should be carried be solar and how much by auxiliary energy heater.
- The temperature requirement in the space i.e. temperature at which heat would be available and used for space heating.
- Temperature stratification, if any
- The means of controlling and their costs.

8.10.3 Solar air system for active approach

The solar air system for active heating of building essentially consisted of a solar air collector (for heating air), a storage unit (generally rock bed type), auxiliary system (conventional energy based air heater), automatic air dampers (for directing air flow), air handling duct, blowers and pumps, control unit with necessary sensors, etc. The working of this system includes heating of air through solar collectors and storing hot air in the rock bed type storage and then incorporating a suitable heat distribution system for heating the space as per requirement in following modes: -

- Heating the space directly from the hot air supplied through solar air collectors during day hours.
- Heating the space from the storage unit during late hours or when sunlight is not available.
- Heating the space partly through direct contact of solar air collector and rest of heat through auxiliary air heater during day hours.
- Heating the space partly through storage heat and rest from the auxiliary air heater, when sunlight is not available.
- Heating the space from the complete stand by mechanisms operated through auxiliary air heater only.

The air-based system offers many advantages as compared to liquid based system, these are summarized as follows:

- Only one medium (preferably air) may be used for collecting, storing and transferring the heat into space, this would yield higher overall heating efficiency.
- Air systems are more durable, convenient and reliable. The efficiency of air system is also better as compared to other working medium.
- The combination of thermal and operating parameters of air like specific heat, flow rates, density, low inlet temperature can yield higher collector outlet temperature.
- The problem of corrosion is eliminated in air heating system.
- Air collectors are more economic and air leakage is not serious.

- The mechanism involved in control unit is simple and it is readily available.
- There are no freezing and boiling problems as water-heating system has.

Though most of space heating systems uses hot air for operation, but along with its merit, the air heating system involves few disadvantages, which are as follows:

- It involves relatively high pumping cost, because large volume of air is required in heating.
- Since large volume of storage system is required, therefore relatively large volume of storage system is required. Hence more cost involvement as compared to other system of heating.
- The combining of solar air heating system with auxiliary air heater (preferably air conditioning system) is quite difficult.

Various types of solar air heater are used in active heating system of building, few commonly used are as follows :

- Simple conventional air heater.
- V-corrugated type air heater
- Finned type air heater
- Overlapped glass plate air heater
- Matrix type air heater
- Porous bed air heater.

8.10.4 Solar liquid system for active approach

The solar liquid system consists of heating the liquid through solar collectors. The mode of operation of solar liquid system is similar to solar air system. The liquid heating system consists of liquid flat plate collectors, storage systems, and auxiliary system, radiant or connective panels, heat exchangers & other control units. As compared to air heating system, this liquid heating system is having atleast three heat exchangers, one in the collection unit, second in the storage unit and third in room, where heating is done. The water heating system has few advantages as compared to air heating system, such as:

- The water heating systems are widely used and can readily supplied combined space heating and cooling mechanism.

- As an added advantage, this system can supply hot water for domestic applications.
- The liquid collectors operate at higher efficiencies at same collector inlet temperature and can provide domestic hot water efficiently compared to air collectors.
- It requires less space as compared to air collectors.

The disadvantages in liquid system are :

- Freezing & boiling problem in collector water.
- Corrosion problem because of water circulation.
- Hazards due to leakage.
- Higher cost due to additional heat exchangers, leak proof joints and corrosion resistant metals etc.
- Law overall efficiency because of more number of heat exchangers.
- Comparatively low durability.

The general operation, mode of action, size of storage unit and heat distribution system is more or less similar to air heating system and it depends on the system heating system and it depends on the system heating load and geographical locations. There is various types of liquid collectors, some of them are: -

- Corrugated sheet type flat-plate collector
- Roll-bond type of flat-plate collector.
- Tube-in plate type of flat-plate collector
- Water trickle type collector.
- Evacuated tube type collector.

8.10.5 Solar heat pump type active houses

The heat pump technique is based on mechanical engineering gadget, which provides heating or cooling by using a reversible refrigeration cycle. Essentially a simple heat pump consists of following components (Fig. 8.13)

(1) Compressor

(2) Condenser

(3) Evaporator

(4) Working fluid (Freon)

(5) Expander

Basically, the heat pump in heating mode attracts heat at low temperature from outside air and rejects heat at higher temperature to the room air. Where as in cooling mode, a reversing value reverse the roles of evaporator and condenser thus heat is extracted from indoor air providing cooling and rejected to the outside air. In case of solar assisted heat pump, the coefficient of performance increases because the source temperature increases.

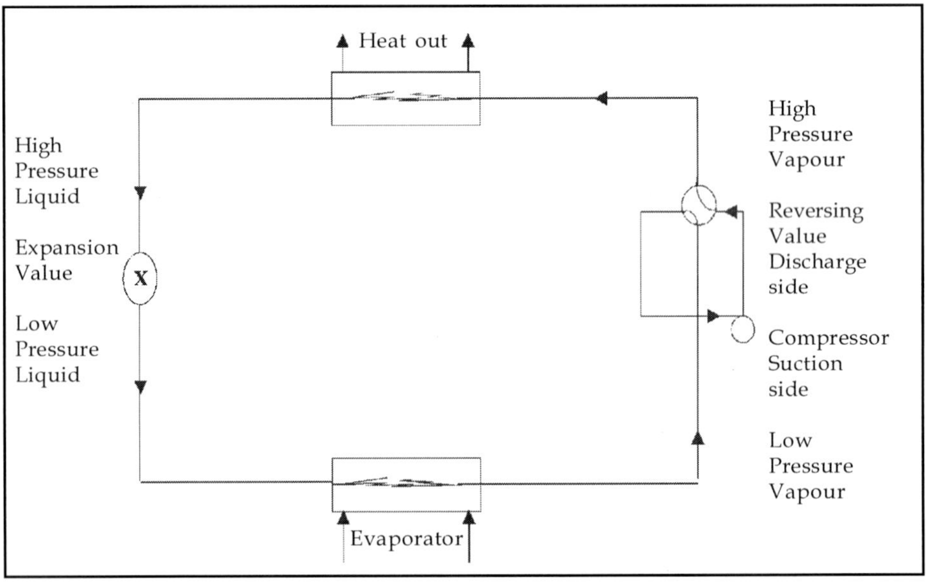

Fig. 8.13 Operation of Heat Pump

In the solar assisted heat pump the working fluid is in the gaseous state, which is compressed and send to the condenser, where it gets cooled to liquify and releases large amount of heat to the surrounding. Then, through expansion value, this liquified working fluid gets vaporized. The evaporation in the evaporator requires large amount of heat, which is drawn from the surroundings. This large amount of low temperature heat is supplied to the evaporator by solar heated air or water. In this system by using appropriate value and controls the heating in winter and cooling in summer can be provided by the same unit. The solar assisted heat pump can be of following types: -

- Air-to-Air type.
- Air to Water type
- Air to Earth type
- Placement of evaporator and condenser respectively.

8.10.6 Solar active cooling techniques

The "active" features are usually incorporated, since they make possible to control the internal climate and heat distribution more precisely. The active solar systems use solar panels for heat collection and electrically driven pumps or fans to transport heat/cold to the living area or to the storage. Electronic devices regulate the collection, storage and distribution of heat within the system.

Active solar cooling is defined as utilization of sun's energy to help offset net cooling load of space conditioning and refrigeration of a building. Many options are available in this direction, which include absorption cycle, desiccant and solar driven "heat engine" cooling etc. However, some mechanical assistance is also required for operating the system.

There is also wide spread interest in desiccant cooling because desiccants can be combined with direct or indirect evaporative coolers to eliminate all refrigerants in the cooling cycle. As in the present context, everyone is aware of global warming and chlorofluorocarbon issue, therefore it is believe that desiccants will become more popular, as regeneration of the desiccants can be accomplished with typical flat-plate collector and neither desiccant nor absorption cooling systems use CFCS. It also eliminates chances of carbon dioxide emission, which contribute to global warming.

Since cooling is required in the summer when the sun is shining, or when heat is available in the atmosphere. Because of this advantage, the solar refrigeration (preservation of products) and air conditioning (cooling the space) required a significant position in present context.

The refrigeration and air conditioning using solar energy can be achieved by one of the following means:

- Vapour compression systems using heat engines or photovoltaic system driven electric motors.
- Vapour absorption system using liquid absorbents such as LiBa, H_2O, H_2O, -NH_3, LiCl - H_2O, NH_3 - $LiNo_3$, R22 - DMF, NH_3 -

NaSCN, or using solid absorbents such or $CaCl_2$ - NH_3, silicagel - H_2O, Zeolites-H_2O.

- Vapour jet system
- Evaporative cooling system
- Thermo-electric cooling system.

A solar air conditioning system consists of following components:

- Field of solar collectors (simple flat plate collector or evacuated tube collector or concentrating collector) depending on the temperature requirement.
- A heat storage device, which gets heat from solar collectors and is used to operate the cooling device.
- A solar cooling device (based on Absorption or Rankine Cycle).
- A cold storage device.
- A heat rejection device.
- Air handling system.

Operation wise, the field of solar collector heat the heat transfer fluid, which is used to operate the cooling device. A part of the heat can be stored in the storage unit. The heat collected from the building is rejected to the atmosphere using a tower or any other suitable heat-rejecting device. If air is cooled by the cooling device then it is directly supplied to the building to be cooled or if chilled water is produced then it is circulated through fan coil units and a part of a chilled water is stored for use when the cooling device is not in operation. The schematic of solar cooling system is shown in Fig. 8.14.

(a) Solar Driven Vapour Compression System

The conventional Vapour Compression System (VCS) can by converting solar energy into mechanical energy through one of the following means.

(i) Heat Engines

Which convert heat energy into mechanical energy or work of air conditioning. The operation of heat engine can be governed through various thermodynamic cycles such as Rankine cycle, Brayton cycle and Stirling cycle. In order to get low temperature Rankine engine is most appropriate. The working of Rankine engine is similar to the steam engine or engine using organic vapour. The use of water vapour (steam) driven Rankine cycle based heat engine is used for high temperature range above 250°C, where as some type of Ranking cycle based heat engine consistent with flat plate or line focussing collectors of low concentration ratios is suitable for low to medium temperature applications (between 90 to 150°C). The Stirling cycle and Brayton cycle based heat engine driven vapour compression type of air conditioning unit is suitable for high temperature application. The operation of heat engine based on different thermodynamic system is shown in Fig. 8.15.

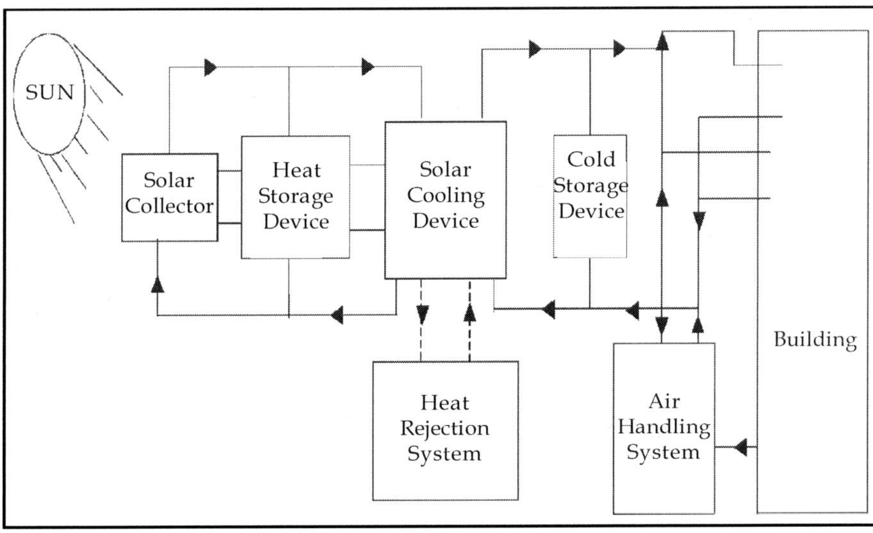

Fig. 8.14 Solar Cooling System

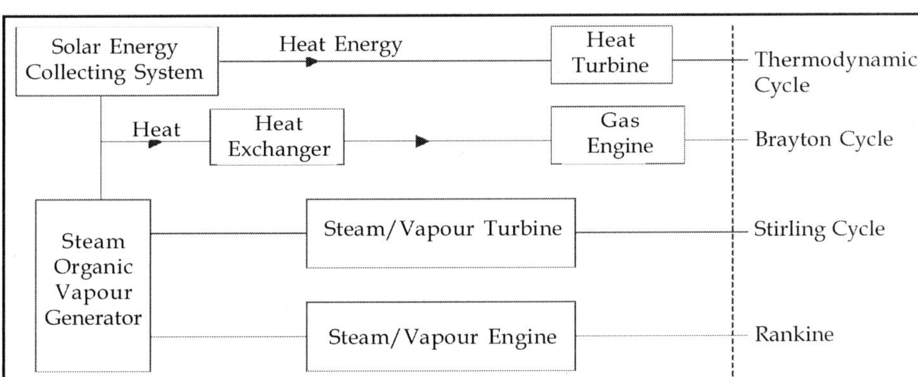

Fig. 8.15 Solar Driven Vapour Compression System

The solar Rankine vapour compression system based air conditioning systems are available with different capacities. The overall coefficient of performance (which is the ration of the amount of cooling to the energy input) for a Rankine cycle operated solar cooling system is of about 0.3 to 0.4. The COP of Rankine cycle depends on the efficiency of solar collecting field. The solar Rankine vapor compression cooling process can be used in the heat pump mode also, and for electricity generation as well when cooling is not required. However, controls of the system need attention during variable solar insolation.

(ii) Solar Photovoltaic Based Electric Motor Operated Vapour Compression System

The vapour compression cooling process operated by photovoltaic panels gives a coefficient of performance in the range of 0.25 to 0.35. It may be because solar cells have lower efficiency. This system can also be used in the heat pump mode and for electricity generation when cooling is not required. However, it is very costly system, it is due to high cost of solar cell.

(iii) Solar Operated Vapour Absorption System (VAS)

This type of system requires very little quantity of solar energy as compared to vapour compression system. Therefore, it is a prospective device for exploiting solar thermal energy for producing cold. Essentially it includes two working fluids i.e. refrigerant and absorbent. The selection of absorbent and refrigerant is so made that the absorbent has a high affinity for the refrigerant. Essentially absorbent ? refrigerant solution is liquid. In this type of system, solar energy is used for heating a strong solution, which is rich in refrigerant. The heating is through solar collector and such that the vapour pressure equals the saturation pressure in the condenser. The refrigerant in vapour form goes to the condenser, while the weak solution returns to the absorber through a recuperator and a throttling value. In the condenser the refrigerant gets condensed rejecting heat and comes in the liquid form at high pressure. The refrigerant now passes through the expansion value and evaporates in the evaporator, thereby taking heat from the surrounding. Thus the cold air goes out from the evaporator and cools the space. After that the low vapour pressure refrigerant from the evaporator goes to the absorber where it is reabsorbed with the liberation of heat. The low pressure, rich reabsorbed refrigerant solution is now pumped to the generator at high-pressure to complete the cycle.

In this system the evaporator and absorber of the system are in the low-pressure side and the generator and condenser are in the high-pressure side of the system. A recuperator heat exchanger is also used to transfer heat between the solutions passing between the absorber and the generator.

The operating principle of absorption cooling in shown in Fig. 8.16.

The vapour absorption system suitable for solar energy operation can be classified as follows: -

- Intermittent cycle
- Continuous cycle - closed and open type

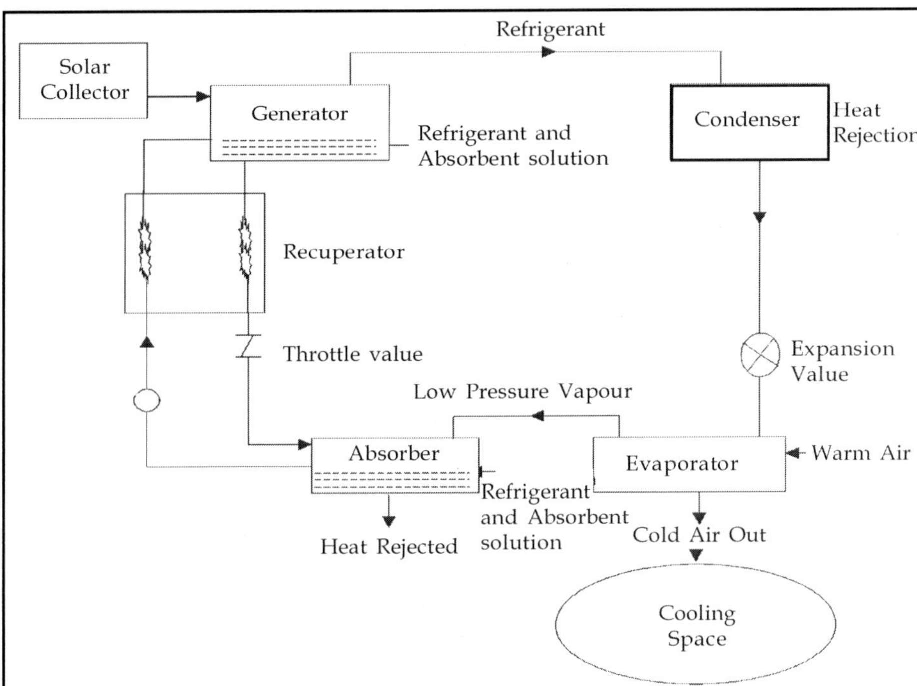

Fig. 8.16 Principle of Absorption Cooling System

The coefficient of performance of the cooling system based on absorption closed cycle is 0.1 to 0.2 depending on the collector efficiency.

Intermittent vapour system especially developed for operation with solar energy take into account the intermittence due to daily variation of solar radiation. In this system, ammonia as refrigerant and water or NaSCN as absorbent and Freon 22 as refrigerant and DMF as absorbent can be used.

This system yields low coefficient of performance because high heat losses are associated with alternately cooling and heating of the unit. As a result of it large physical size of the system is required for a given capacity. Further, as the nature of this system concerned, it is not capable for producing cooling on continuous basis.

However, the intermittent VAS is simple in design and construction and it does not require refrigerant pump. The operation of this system is shown in Fig. 8.17.

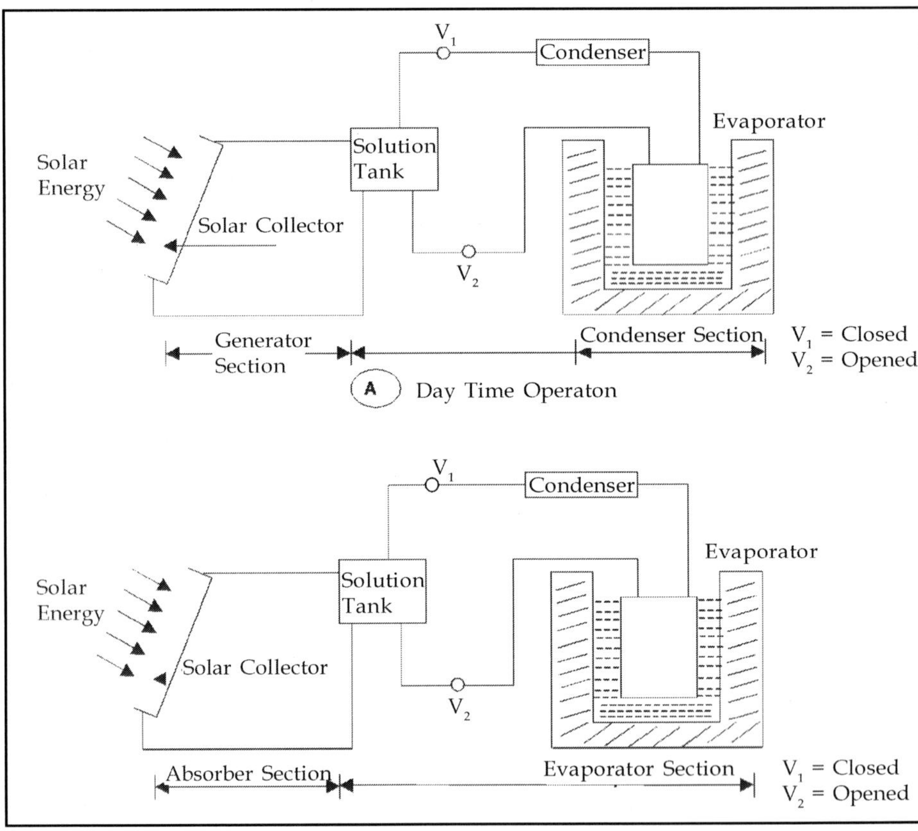

Fig. 8.17. Principle of Intermittent VAS

In a continuous vapour absorption system, heat can be recovered continuously by using a regenerative heat exchanger to transfer heat from the hot weak solution to the cold strong solution. In case of closed cycle continuous vapour absorption system, method of supplying heat of generator. In such integrated collector- generator systems, it will be necessary to have storage reservoirs for weak solution, strong solution & the liquid refrigerant.

The open cycle continuous VAS is similar in operation to the closed cycle except that the weak solution is regenerated by losing the refrigerant to the atmosphere in place of the refrigerant being recovered in a condenser.

(iv) Desiccant Cooling

Dehumidification is also a method of cooling. In fact, dehumidification is a method of removal of moisture from the room air. This can be carried by using absorbent followed by evaporative cooling of air. This is very promising way of active cooling for hot and humid climate. There are few liquid & solid materials, which have property of attracting and holding water vapour known as desiccants. These materials can be used for dehumidify the air and can thus be used for cooling the building. These desiccant may be absorbents or adsorbents type. The desirable characteristics of desiccant are:

- They should be thermally and chemically stable.
- They should be non-toxic, non-flammable, odorless & non-corrosive.
- Their cost should be minimum, as far as possible also they should be readily available.

Both liquid and solid desiccant can be used in open as more as in closed cycle.

8.10.7 Passive solar architecture heating

In a passive solar heating system, the solar energy collection, storage and distribution is done by natural means and for this no external means either electrical or mechanical means are used. In designing and constructing buildings in order to take advantage of the Sun's energy and the sub-sounding environment. The solar passive building itself collects stores and uses solar energy and is designed to take advantage of site specific features for natural heating, cooling and lighting.

In this system, various building elements such as walls, roof, windows, partition etc. are so architecturally selected and arranged that they contribute effectively in solar energy collection, storage, transportation and distribution of thermal energy as per requirement. In passive solar house, heating is done by natural on-site energy sources such as solar radiation, outside air and internal metabolism, while cooling is meet out through sky and space, outside air and wet surface. Both direct as well as diffuse solar radiation is used for heating the house. Apart from this, building elements such as construction materials like stone, bricks, concrete, water insulation, glazing, shading, reflectors etc. and various thermal processes such as thermal radiation, natural & forced convection, conduction, air stratification, evaporation, themosyphoning etc. are combined in various ways to get desired heating or cooling.

(a) Solar Passive Architecture - Conceptual Details

Solar passive architecture incorporates several features such as shape and orientation of building, shading devices and use of appropriate building materials for conserving energy used in heating, cooling & interior lighting of building considering prevailing solar radiation conditions in the area.

There are few important points that are kept in view before drawing up building plans. Orientation of building in right direction to get the maximum benefit of natural ventilation and lighting as it reduces demand for artificial lighting and also provides natural cooling and heating is one of important consideration for proper building plans. The proper orientation of building must includes following points:

- The living room, the bedrooms and the kitchen must be windy and naturally ventilated.
- The wind must blow into the living room from the south/ south-west direction through the entrance door, if possible.
- A French door (high door) may be oriented in the east and designed to allow sunshine into the room in winter but not in the summer. Thus the design and position of door must takes into account the movement of the sun in all seasons.
- Care must be taken to see that neither the living room nor any of the bedrooms have a wall that directly faces the west.
- The roof ceiling is also at a sufficient height in the living room so there is enough space for the hit air to rise and the living area near the floor remains comparatively cooler.

- A light well may be incorporated in the house, which keeps the home fresh and cool, it is a source of natural light that keeps the inside of the building bright throughout the day. If possible, it should be provided right at the building bright throughout the day. If possible, it should be provided right at the centre of the house so that when doors of the living room are opened, cool air from the outside rushes in, driving out the hot air within the house through this light well, which is provided with lowers at the top.
- It is not enough to improve the insulating value of the house. This saves thermal energy but will increase the electrical energy load for artificial light, if it involves an inadequate area of glazing. A proper balance is to made between the benefit of natural lighting and the loss of heat through windows.
- Consideration must be given to the profile of the building, the location of the windows and the depth of the usable floor area in relation to the line of the windows, so that the utmost benefit can be obtained from natural lighting.
- The building envelope must be tight, so that the leakage of unwanted air from outside can be eliminated. The windows are placed in such a manner that they must be open able in order to give relief to the occupants by the provision of adequate natural ventilation.
- Building mass should be properly deployed so that it can be useful for storing unwanted solar heat during peak hours for use at a later tine when they can offset a heat loss.
- The location of thermal insulation in a building is quite important when it is placed on the inside of the house it minimize the effect of the building mass and when placed on the exterior it increases the effect.
- If possible, a basement is provided under the living room. Three fourths of the basement is underground and one side should be partially exposed to get fresh air. This keep the basement cools and ventilated. As there are enough natural lights and the earth embankment on the three sides keeps it cool, the basement is a favorite place to work or relax on hot summer afternoons.
- External shading should be arranged to minimize unwanted solar gains in summer.

- The material chosen for building construction should be reviewed in relation to the energy used in their manufacture. As an example, reinforced concrete beams appear to need less energy for their manufacture, transport & erection on site than do steel beams. Similarly brick seems to be more economical than concrete, in terms of energy & timber windows frames better than aluminum.
- Growing ample greenery around a concrete structure (house) is also a passive solar feature. Green plants absorb heat and offer shade to the walls and the roof from the direct assault of the sunrays. Plants and trees, because of their dark color, large structure area and evaporative cooling, don't reflect heat towards a building nor store heat for later re-radiation to the building.

Thus a complete passive solar heating system includes collection of energy through south facing glass, storage of energy through the use of thermal mass (usually in the form of concrete, brick, water or phase change material), regulation of energy through provision of overhangs, shades or other insulating material for windows or providing a door to close off a sun space at night and distribution of energy, which usually consists of vents, dampers, duct work and small fans. Therefore the design of houses according to passive approach requires a detailed understanding of the complex inter-relationship between architectural textures, human behaviour and climatic factors. By arranging local features such as vegetation, topography & solar exposure, the climate in the house can be favorably altered to improve thermal comfort. Apart from heating and cooling, there are many other factors, which need to be considered for one comfort levels These are relative humidity, air flow and mean radiant temperature (temperature radiated from objects in the living area). If the humidity in room is low, the air movement high and the walls & floors of the room become cold in nature & as a result of it we feel cool inside the room even when actually the temperature is high. But if the humidity is high, air movement low and radiant temperatures high and as a result of it, inside of room become warmer, even if the temperature is low.

(b) Types of Passive solar heating systems

The passive solar house are classified as:

- Direct gain type
- Indirect gain type (thermal storage wall and thermal storage roof)
- Isolated type gain system (attached sunspace type & convective loop type).

(i) Direct Gain Type

This is one of the most common type approaches, which include capturing the sun's rays directly by using large south facing window and glazed surfaces on the walls and on the roof.

The placement of glazing over the window facing south or the entire south-facing wall permits solar radiation entry inside the room. Almost all the solar radiation enters in the room is converted into heat energy (Fig.8.18).

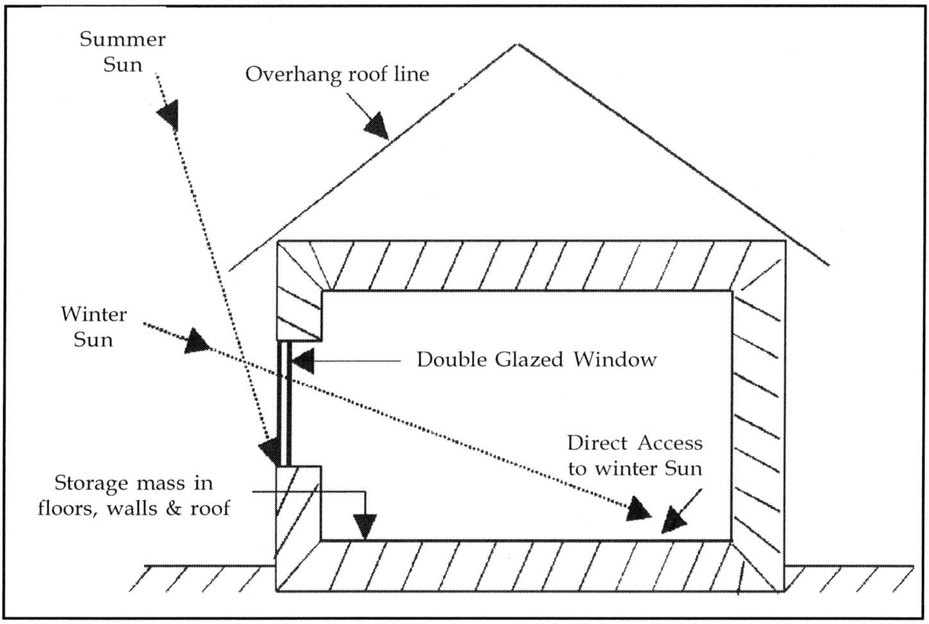

Fig. 8.18 Direct Gain Type

The heat loss from the room is reduced by using a double glazed window. An appropriate overhang above the window is provide which only permit winter sun to enter, where as summer sun is not allowed to enter inside the room. The seasonal variation reflects the position of sun, during summer the elevation of the sun is high (Fig. 8.19).

A sufficient thermal mass in the form of massive floor area and wall is required, which absorb sunlight during day hours and reflect heat during night hours. Some times, insulation, over the window is also required which is used during night, when heat loss is more than heat gain through windows.

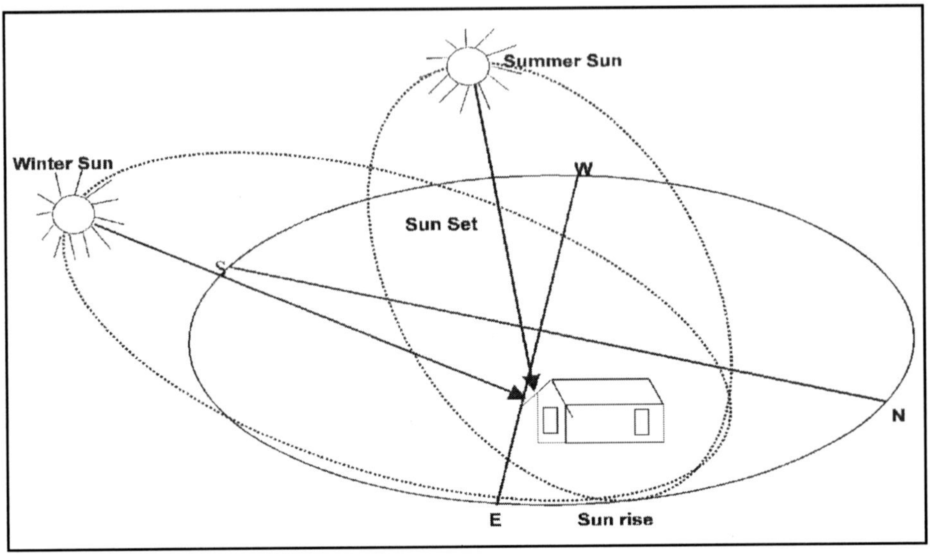

Fig. 8.19 Position of Sun According to Season

(ii) Indirect Gain Type

Idea behind this type of approach is similar to that direct gain type solar houses, except that heat is collected in some special architectural feature such as thermal mass is placed between the glazed solar collection area and the interior space and thus solar radiation is blocked from entering directly into the room to be heated. Here the heat is transferred indirectly from the storage mass by natural convection and radiation. The heat transfers from the storage mass either wall or roof is continuing as long as its temperature is above that of the inside temperature of the rooms.

Indirect gain type solar houses provide effective way of heating the houses and reducing the large oscillations in the room air temperature. Essentially thermal storage wall of indirect gain type solar house may be constructed of masonry or concrete materials or formed by the use of water filled containers. The masonry or concrete wall is known as trombe wall where as water filled containers type of wall is known as water wall. The schematic of trombe wall and water wall is given in Fig. 8.20.

Operation wise, when solar radiation passes through the glazing and strikes the wall and as a result of it heat is absorbed by the dark exterior of the wall. After certain duration, temperature of wall rise and than heat begins to flow into the interior of the building. This time period is called time lag of the wall. Which depends on the thickness and the material of the thermal storage wall. Also, there is air gap between the wall and glazing surface. In this air gap, hot air moves from bottom to top generally due to natural convection. This heated air goes in the room through the upper vent while the cool air from room through the bottom sent enters in the gap. This circulation is continuing till the wall goes on heating the air. In case of peak summer, when the temperature of air inside the air gap increases to large extent, in this case upper dampers provided at the top of glazing surface, opened outward and thereby excess heat can be removed.

Such a thermal storage wall made of masonry and concrete is known as Trombe wall, while the thermal storage wall made up of drums or barrels of water is known as water wall. The water wall, which consists of barrels of water, which, stacked over each other, collects, stores and distributes the heat into the room. An insulating sheet is also provided on the glazing surface, which can be opened and closed as per requirement of the heat entry inside the room. During sunshine hour, the insulating layer is removed, which allows solar energy to enter inside the room, while during off sunshine hours, insulating layer is placed over the glazing to prevent back flow of heat from the room to outside.

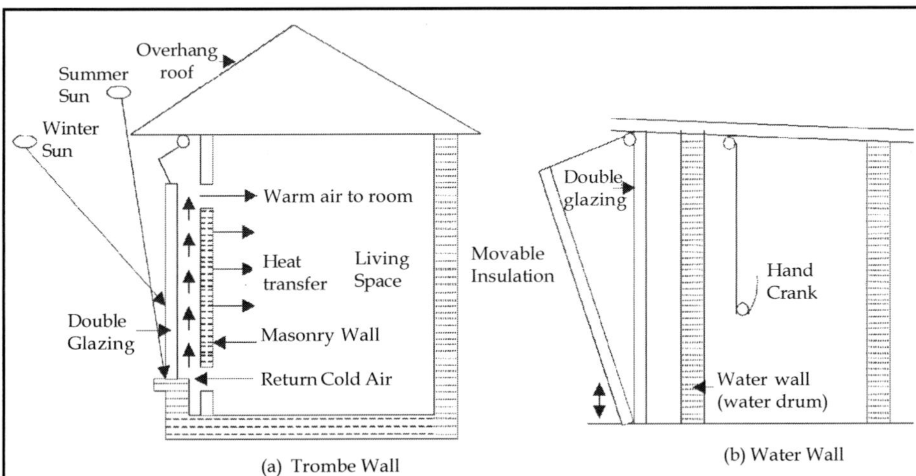

Fig. 8.20 Indirect Gain Type Solar House

Thermal storage roof is also reflects indirect gain type of system. The basic difference between thermal storage wall and thermal storage roof is that in the storage roof case the interposed thermal storage mass is on the building roof instead of a wall generally, a metal roof is used for conducting heat effectively. Over the roof, water bags made of transparent or black plastic sheet filled with water or any other massive materials is placed, which used to store heat during day hours and heats the room below it during day hours as well as during night hours or whenever heat is required. A provision of movable insulating shutters over the water bags is also provided, which direct heat supply to room, whenever it is required.

(iii) Isolated Gain Type Solar House

This is a system of passive house in which direct as well as indirect gain concept is integrated to get full advantage of passive technology. It may be attached greenhouse type or convective loop type. In case of attached greenhouse type or convective loop type. In case of attached greenhouse type a additional sunspace is provided between the interior space (where heating is done) and the solar heating glazing surface. This sunspace is on the extreme south side, which can be used as green house for raising vegetables or flowers. It has been observed that this attached sunspace has large air temperature swing.

The temperature is quite high during summer. The heat gain in the greenhouse is imparted to the thermal wall, which is in between room (where heating is required) and green house. The heating is conveyed through radiation and convection made inside the living room (Fig. 8.21).

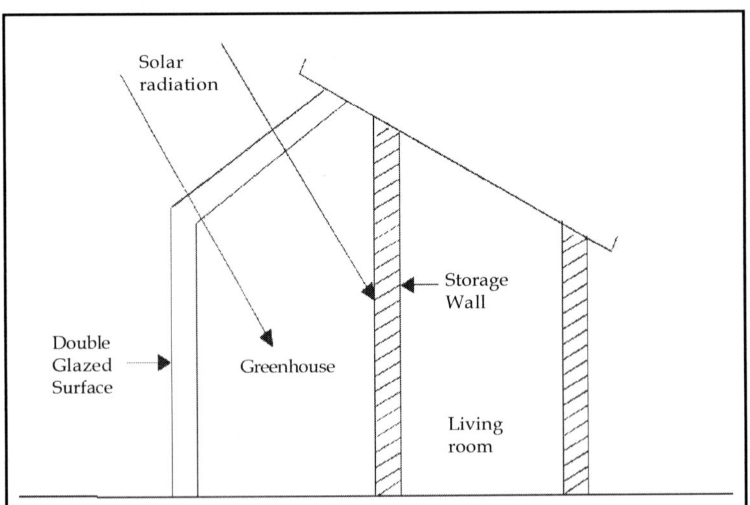

Fig. 8.21 Isolated Type Solar Passive House

Convective loop isolated type solar house similar to the general characteristics of active system, in which a separate solar collector & thermal storage bed is provided. In this system, air after getting heated through solar air collector goes directly into the living room for heating or through a rock bed storage unit for heating the storage unit when heat is required vents are opened to the thermal storage to allow warm air to flow into the room by convection (Fig. 8.22).

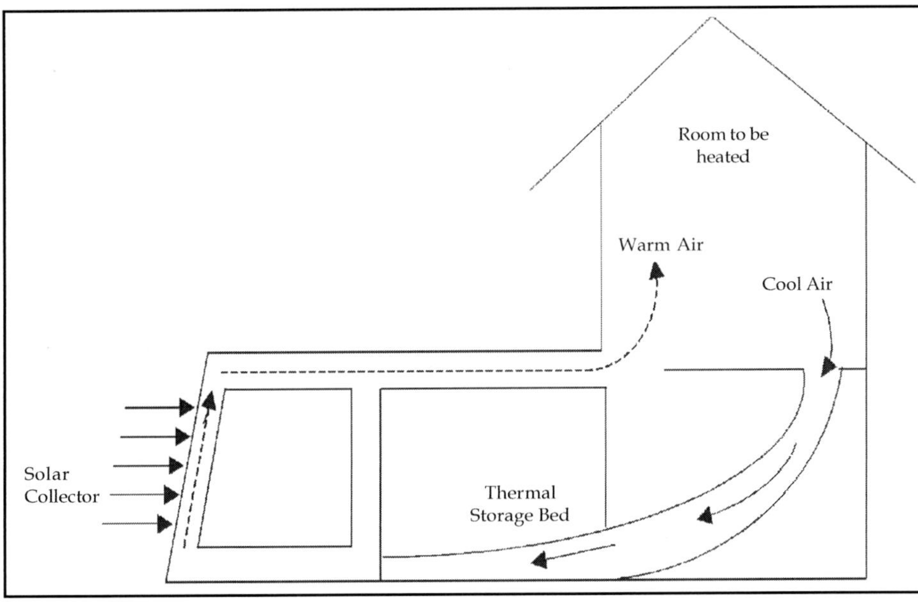

Fig. 8.22 Convective Type Solar Passive House

8.11 Solar Ponds

With its vast potential of virtually untapped energy, living off the sun promises to be an attractive for the future. However, because of the intermittent and diffused nature of solar radiation, the problem of the high cost of collecting and storing solar energy neglects the economic viability of large-scale solar thermal applications. For such applications, it is more effective and economical to devise a solar system that combines both the collection and storage functions in place of the conventional flat-plate solar collector and hot water storage tanks. Large expanses of water bodies like ponds, lakes or oceans are natural potential solar collectors with built-in storage characteristics. Experience shows that the water in such a pond heats upto only a few degrees, because of the natural convection currents that are set into medium as soon as heat is absorbed at the bottom of the pond.

In a mass of water of low depth, the solar radiation falling on the water will penetrate and be absorbed at the bottom, raising the water temperature. But buoyancy will immediately cause this heated water to rise to the surface and the heat will be rapidly dissipated to the surroundings. If the water in the lower regions of the pond of lake could be made heavier than at the top, it could stay at the bottom and retain the absorbed heat, thereby yielding larger temperature differences between the bottom and the surface layers of water.

This effect has in fact been observed in some natural lakes whose water is found to contain a concentration of salt that increases with the depth. The resulting 'salinity gradient' (salt-slope) gives the lake a vertical density difference so that heated 'heavy' water remains at the bottom. A significant temperature rise of 40°C to 50°C has been observed in such waters.

Solar ponds is an artificially constructed water pond in which significant temperature rises are caused to occur in the lower regions by preventing convection currents. The more specific terms "salt gradient solar pond" or "non-convecting solar ponds" are also often used. At the moment solar ponds are mostly being studied on an experimental basis in a number of places. However, they appear to be economical for large area applications and it is likely that they will be used extensively in the future as problems connected with their operation and maintenance are resolved.

The solar pond, which is actually a large area solar collector, is a simple technology that uses water pond between one to four meters deep as a working material for three functions: *(a)* collection of solar radiant energy and its conversion to heat energy (upto 95°C), *(b)* storage of heat and *(c)* transport of thermal energy out of the system. It possesses a thermal storage capability spanning the seasons. The surface area of the pond affects the amount of solar radiation it can collect. The bottom of the ponds is generally lined with a durable plastic liner made from material such as black polyethylene and hypalon reinforced with nylon mesh. This dark surface at the bottom of the pond increases the absorption of solar radiation. Salts like magnesium chloride, sodium chloride or sodium nitrate are dissolved in the water, the concentration being highest/densest at the bottom (20 to 30 per cent) and gradually decreasing to almost zero at the top (Fig. 8.23).

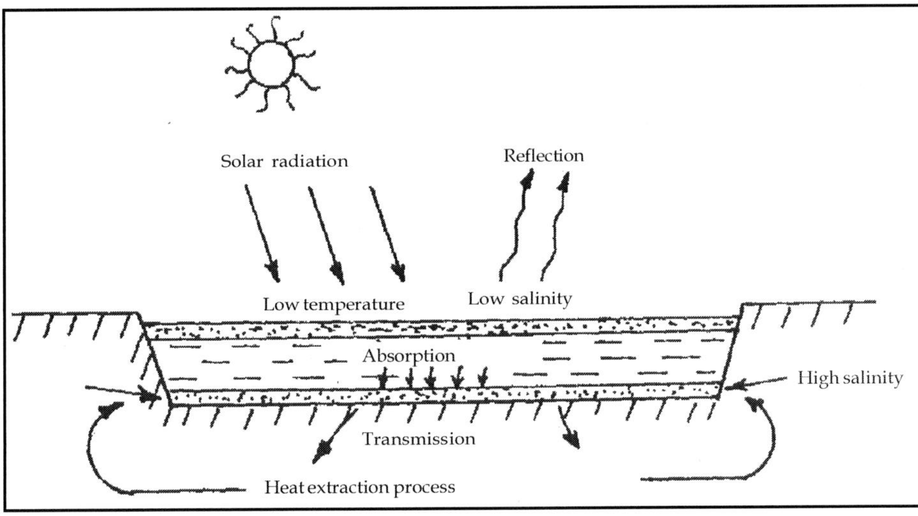

Fig. 8.23 Cross Section of Solar Pond

Typically a salt-gradient solar pond consists of three zones: *(a)* an upper convective zone of clear 'fresh' water that acts as the solar collector/receiver and which is relatively the most shallow in depth and in generally close to ambient temperatures, *(b)* a gradient which serves as an insulating layer by reducing heat losses in the upward direction. This non-convective zone is much thicker and occupies more than half the depth, and *(c)* a lower convective zone with the densest salt concentration, which serves as the heat storage zone. Almost as thick as the middle non-convective zone, salt concentration and temperatures in this zone are nearly constant. The Solar Pond is filled in stages. To obtain the difference in salt density through this stepped concentration gradient, the pond is filled with atleast 3 successive layers of salt solution (10 to 20 cm thick), one on top of another, each less dense than the layer below, so that the top layer is fresh water or nearly so, while the bottom layer contains the most salt. Naturally this kind of stepped concentration is not likely to remain stable for long, and would eventually disappear because of diffusion and evaporation. In order to maintain the stability, concentrated bring is introduced at the bottom periodically while the surface is "washed" with fresh water. If the pond contained no salt, the bottom layer (which is denser in the solar pond) would be less dense than the top layer because hot water expands. The less dense hot water would then rise to the surface and the layer would mix. Therefore the purpose of the salt density difference is to keep the different 'layer's of the solar pond separate.

When solar radiation strikes the pond, most of it is absorbed by the surface at the bottom of the pond. The temperature of the dense salt-water layer is therefore higher than the fresh water layer near the surface of the pond. The denser, salt water at the bottom prevents the heat being transferred to the top layer of fresh water by natural convection currents.

In order to extract the energy stored in the bottom layer, hot water is removed continuously from the bottom, passed through a heat exchanger and returned to the bottom to generate electricity heat stored in the salt water is piped to an evaporator. Liquid freon in the evaporator is heated and changed into a gas. The pressure generated by the gas spins a turbine and electricity is produced from the generator. Freon gas is then cooled and recycled to be used again.

8.12 Solar Thermal Power Generation

In the solar thermal power generation route, solar energy is collected on a collector's surface and then transferred to an absorber. A fluid flowing through the absorber transfers heat to a heat power generating machine i.e. turbine. Where heat energy is converted into mechanical energy and in turns a generator is attached with turbine where mechanical energy is converted into electrical energy. The thermodynamic heat cycle here works exactly in the same way as in the case of conventional steam based power station. Here only difference is that the heat source is absorber of the solar collector instead of a boiler or a combustion chamber. Two types of solar thermal power generation systems are available:

(i) Centralized system (tower concept)

(ii) Distributed system (farm concept)

In the tower concept, a field of heliostats, which follows the direction of sun, concentrates direct solar radiation into a central receiver mounted on the top of a tower. Fig. 8.24 gives a schematic diagram of a solar thermal tower power plant. A heat transfer fluid removes heat from the receiver and in turn carries it to another heat storage unit. This heat source is used to operate an electrical power plant. The advantages of the farm concept are its modularity and relatively less maintenance.

Here the heliostats consist of glass mirror mounted in segments in a steel frame. The mirrors are backed by formed core / steel sheet laminates. The heliostats are usually supported by stainless steel pipes on reinforced concrete foundation.

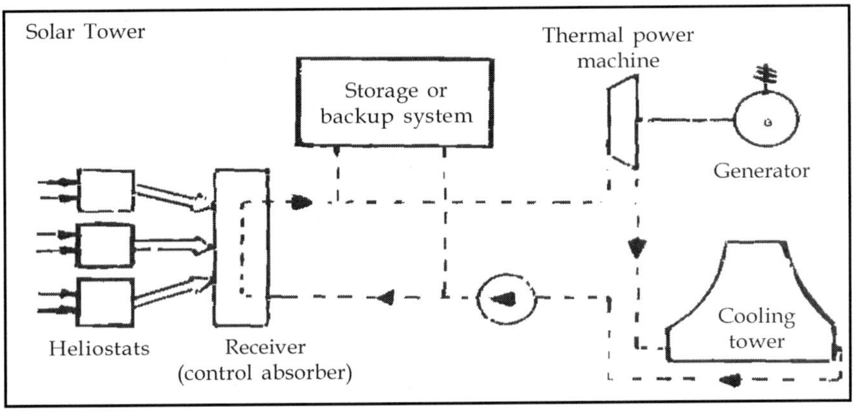

Fig. 8.24 Solar Tower Concept of Electricity Generation

The performance of the heliostats depends mostly on the following parameters:

(i) *Selected mirror construction* : It should ensure that deformation which may result from wind, temperature, etc. remain within the prescribed limits.

(ii) *Surface treatment of mirrors* : To maintain the reflectivity constant during their lifetime.

(iii) *Stability of the supporting and tracking mechanism* : It is used for frequent automotive adjustment of the mirrors according to sun.

The other important component in the receiver, which is usually designed on a cavity to minimize the radiation losses, is essentially a boiler or steam generator. It is located at the top of the tower, and it receives the sun's energy directed on it by the heliostats. An array of spaced tubes forms the boiler surface. The heat transport media (water, gas or medium depending upon the chosen heat cycle) flows through these tubes. As temperature of 1000°C or more can be reached, the use of suitable materials and its design are crucial. Here on tower concept, higher temperature and hence higher thermal power is possible. It is more suitable for central utilization. Maximum possible power generation from tower concept is about 100 MW.

In a farm concept (Fig. 8.25) flat plate or parabolic dish or trough collectors may be used for generation of electricity. Each collector is connected to the forward flow tubes and the return flow tubes of the central thermal power machine. A parabolic trough collector consists of an axial parabolic reflector at the focus of which a heat receiver in the form of a tube is located. Parabolic troughs are usually made of glass segments hot moulded to shape, conventionally silvered on the back face. After silvering, the edges are often sealed by a special substance to prevent moisture and other particles from entering between the glass and silvered layer and spoiling the mirror effect. The segments are then provided with a backing surface, which could be metal, fiberglass or concrete. Concentration ratios of between 10 and 100 can be achieved by such collectors resulting in heat transport medium temperature of 150 to 350°C. The reflectance surface is generally mirrored to achieve reflection of the order of 87 per cent. The focal line of the concentrate is usually made to coincide with the axis of rotation of the supports. A number of troughs are combined together to form module.

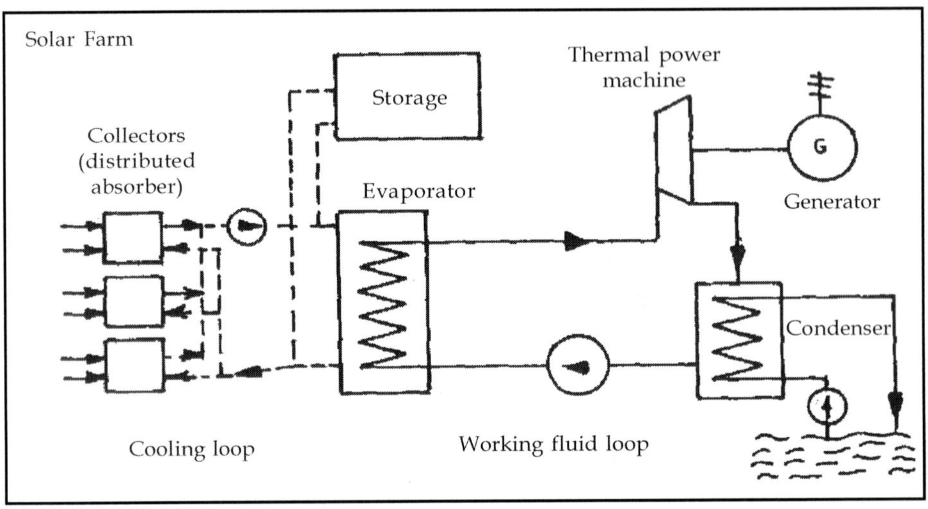

Fig. 8.25 Solar Farm Distribution System for Electricity Generation

A parabolic dish or point focusing solar collector is capable of achieving concentration from about 50 to several hundred, as a result of which temperature of 500°C or more can be reached in the central receiver. A parabolic dish is primarily a parabolic formed by rotation of a parabola along a vertical axis. In other words, cross section of parabolic dish in any direction will reveal a common parabola. Parallel sun's rays falling on the dish are focused on a point and as a result of which high temperatures can be reached at the focus. A receiver placed on the focus could be used in a prime mover for generation of electricity. In parabolic dishes the achievable temperature may range from 300 to 1600°C. Parabolic dishes may also be used for producing industrial process heat. A number of parabolic dishes can be coupled together to generate the required power.

Following types of heat engine are often used with parabolic dish system:-

(i) Organic Rankine cycle engine
(ii) Stirling cycle engine
(iii) Air- Brayton cycle engine

In each of these cycles hot gas or vapor is expanded through an engine or turbine to produce work and is thereby cooled. The gas vapor is further cooled to reject heat and is finally returned to its initial state for getting energized by solar radiation and then complete the cycle.

In this distributed solar thermal power generating system, a hybrid concept has also been developed. In this approach, the working fluid in heated by solar energy to the maximum possible extent. Generally parabolic trough collector systems are being used in this hybrid concept.

In order to maximize the efficiency of the solar system and to ensure steady output during fluctuating solar insulation, a supplementary boiler that can be fired on multi-fuels such as biomass, etc. may also be added.

8.12.1 Design considerations in solar thermal power plants

Electricity generation by use of solar energy through thermal route has been considered an important means of conversion of solar energy to usable form. Solar thermal power plant has different design consideration as compared to conventional power plant. All subsystem should be carefully integrated or method for getting maximum efficiency of generation and utilization. Following points are properly considered for design of solar thermal power plants.

(a) Site selection

Main considerations for site selection for electricity generation are good solar insolation level, preferred latitude between 20° and 35°, plain and wasteland having good drainage and having little economic value, preferably away from polluting industries, dry climate, well connected with transport routes, good availability of water for steam generation, mirror washing and cooling tower, etc. To reduce the transmission and distribution losses, it will be desirable that plant be located near the centre and load curve of the area should preferably have a discernible day-peak.

(b) Requirement of land

While designing a solar thermal power plant for medium latitudes (20° to 35°), for maximum collection of solar energy, it is desirable that the distance between the two rows should be kept at approximately three times the aperture of the collectors. On this basis the estimated land requirement per MW works out to around 7 acres. Land area selected should be such as to minimize piping and ensure heat transfer fluid flow balancing in modules.

(c) Capacity selection

Large size solar thermal plants have obviously to be designed to operate in synchronous mode with the existing grid. It may, therefore, be argued that the capacity of the proposed solar thermal power project should be declared on the basis of grid capacity. In case the capacity of proposed project is decided at a level of higher than the local peak demand, part of generation can be fed to the existing grid for the nearby load centers.

(d) Orientation of collectors

It is also very important aspect for electricity generation through solar thermal route. The most favorable orientation of parabolic trough collectors for a solar thermal power plant is obviously North-South axis. This orientation maximized the year-round electricity generation. The electricity generation almost follows the solar radiation curve and not necessarily the load curve. However, it may be possible to orient the collectors in a manner such that electricity generation is increased during off-noon hours though at the expense of perhaps greater reduction in its quantum around noon. This may be desirable if electricity during off-noon is valued far higher than the electricity around noon. Similarly by orienting the collectors along East-West axis, annual electricity generation for latitude around 26 °N is reduced by 15-20 per cent.

(e) Selection of collector area and other components

Because of diurnal and seasonal variation of solar insolation, the collector field sizing at a particular level is of critical importance. Sizing of the collector field at lower insolation levels would mean that around noon hours, a part of the field would remain unused unless the turbine could take overloading. Sizing of the field at higher insolation levels would result in part load operation of turbo-generator set for most of the time.

To take care of yearly variation in solar radiation, the collector field should be sized on the basis of a Typical Meteorological Year (TMY). It is desirable to conduct techno-economic feasibility studies of a solar thermal power plant on the basis of TMY data.

Other important considerations in determining size of the collector field are: frequency of insolation level at a particular hour; cost of collectors; long term marginal cost of power generation for each unit, increment in collector area, etc.

Pumping of the heat transfer fluid through absorber tubes and the heat exchangers is a major energy consuming auxiliary requirement in a solar thermal power plant. Therefore, proper selection of pumps to ensure continuous flow with highest possible efficiency is important.

The primary requirements are to overcome the system pressure drop at required flow rate and high efficiency.

(f) Heat exchangers

A Solar Thermal Power Plant has a number of heat exchangers such as heat exchangers to transfer heat from heat transfer fluid to water/steam (steam generator system comprising economizer, boiler, super heaters and preheaters), surface condenser, feed heaters, etc. The complete heat exchanger system should preferably be thermally and mechanically designed for about 110 per cent of capacity, based on the turbine-generator nameplate rating.

(g) Turbine

Selection of turbine for solar thermal power plant is another critical consideration. It has been estimated that one point drop in turbine efficiency increases collector area by three points, which means highly efficient turbines are desirable.

(h) Back up capacity requirement

Back-up capacity of the system should be optimized on the basis of load curve, duration of fluctuations in solar insolation, etc. If load demand during evenings is lower than nominal plant capacity and fluctuation in solar insolation is high. Installation of a partial back up is recommended. It is advisable to select back-up capacity after careful analysis of cost effectiveness. To minimize the precipitation of carbon and sulphur oxides which may cause mirror and metallic corrosion, use of certain fuels having non-corrosive property and sulphur heavy stock oil (LSHS), etc. is recommended. In case load demand is high during evening and morning along with frequent fluctuations due to grid failure and cloudy weather and good quality fuel for back up is not available or is costly, a thermal storage is recommended.

9

Solar Photovoltaic Technology

9.1 Introduction

Solar energy can be directly converted into electrical energy through photovoltaic route. Energy conversion devices which are used to convert sun light into electricity by use of the photovoltaic effect are called solar cells. The advantages of solar cells are:

1. They are reliable, convenient and durable.
2. It is green, sustainable and pollution free technology.
3. Directly solar radiation can be converted into electricity (conversion of global solar light by flat plate photovoltaic).
4. Maintenance cost is quite low therefore suitable even in isolated and remote area, since no moving parts are involved.
5. Solar cells are quite compatible with almost all environments, respond instantaneously with solar radiation.
6. Considerable expected lifetime i.e. 20 years or even more.
7. No distribution system is required, because they are decentralized source of energy and located at the place of use.
8. No moving parts in energy generation and conversion, portable and easily assembled at any place in a short time.
9. No distribution and transmission losses.

10. They don't create pollution as compared to other conventional sources of energy.
11. Modular nature in which desired capacity can be designed by simple integration method.
12. Ability to function unattended for long periods.
13. Compatible and integratable with any other power sources.
14. Remote monitoring and control feasible.
15. Recyclable material.
16. Large specific power per unit weight, particularly in case of thin film solar cells.
17. Ideally suited for small stand-alone power systems particularly in remote and grid-less areas.
18. Taking life cycle into account, PV is economically viable as also a more favourable solution for many small / medium power application.
19. As a sun based energy production & distribution industry, it is expected to be a significant drive of world economics and generation of job.

However, solar cells, have same disadvantages, which are as follows:-

1. Solar energy has intermittent nature, therefore, can be used only in presence of sun.
2. Present costs of solar cells are very high.
3. Presently no adequate, reliable, efficient and convenient energy storage system is available.
4. The conversion efficiency is also relatively low, which is limited to 25 per cent.

9.2 SWOT Analysis of Solar Photovoltaic System

The SWOT analysis provides an analysis of the strength, weaknesses, opportunities and threats of SPV. This analysis defines the role of SPV in meeting the electricity needs for several applications.

Strengths: (1) Cost effective: in terms of life cycle (2) Reliable (3) Low maintenance (4) Environmentally benign (5) Sunshine is free and SPV is fuel independent (6) Locally generated power- greater energy security and control of access to energy (7) Flexible size-modular construction (8) Transportability-modular construction.

Weaknesses: (1) Sun dependent (sunlight is highly diffused, intermittent and requires backup storage) (2) High initial cost (3) System maintenance (4) Disposal (batteries) (5) Power type - requires inverter (6) Vulnerability (the modular and portable characteristics leave them vulnerable to theft and venations.

Opportunities: (1) SPV is an appropriate and cost effective source of electricity for basic services and amenities in an environmentally sensitive manner (2) Best option for decentralized and day light energy requirement (such as irrigation).

Threats: Cannot stand if low cost environmentally friendly technology is developed.

9.3 Solar Cell

A typical silicon PV cell is composed of a thin wafer consisting of an ultra-thin layer of phosphorus-doped (N-type) silicon on top of a thicker layer of boron-doped (P-type) silicon. An electrical field is created near the top surface of the cell where these two materials are in contact, called the P-N junction. When sunlight strikes the surface of a PV cell, this electrical field provides momentum and direction to light-stimulated electrons, resulting in a flow of current when the solar cell is connected to an electrical load as illustrated in Fig. 9.1

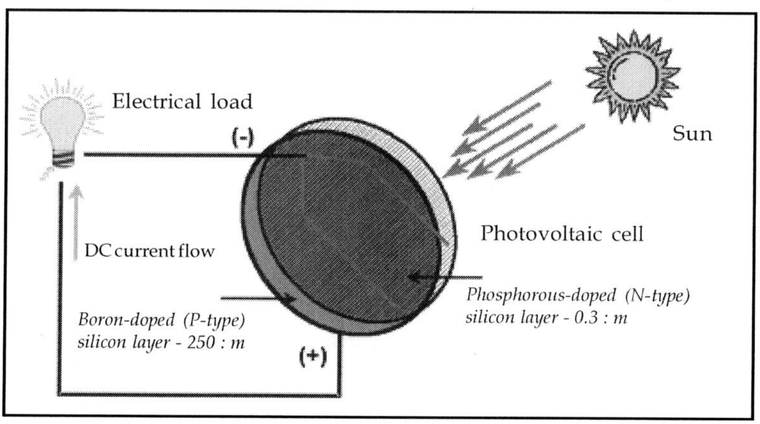

Fig. 9.1: Diagram of Photovoltaic Cell.

Regardless of size, a typical silicon PV cell produces about 0.5 – 0.6 volt DC under open-circuit, no-load conditions. The current (and power) output of a PV cell depends on its efficiency and size (surface area), and is proportional the intensity of sunlight striking the surface of the cell. For example, under peak sunlight conditions a typical commercial PV cell with a surface area of 160 cm^2 will produce about 2 watts peak power. If the sunlight intensity were 40 percent of peak, this cell would produce about 0.8 watts.

9.4 PV Cells, Modules and Arrays

Photovoltaic cells are connected electrically in series and/or parallel circuits to produce higher voltages, currents and power levels. Photovoltaic modules consist of PV cell circuits sealed in an environmentally protective laminate, and are the fundamental building block of PV systems. Photovoltaic panels include one or more PV modules assembled as a pre-wired, field-installable unit. A photovoltaic array is the complete power-generating unit, consisting of any number of PV modules and panels (Fig. 9.2).

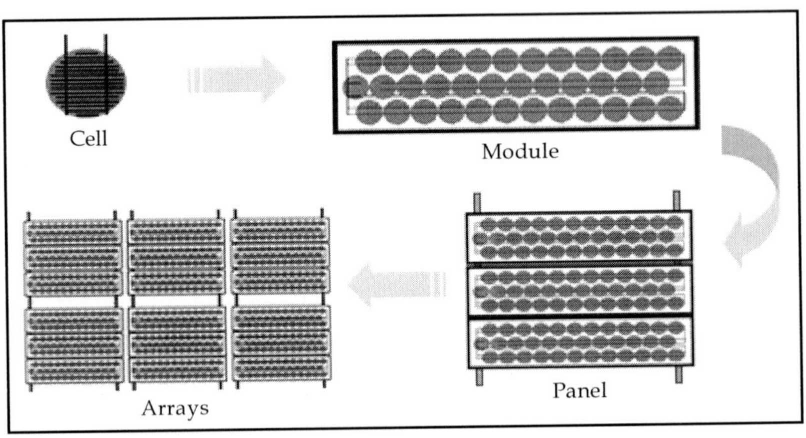

Fig.9. 2: Photovoltaic Cells, Modules, Panels and Arrays

The performance of PV modules and arrays are generally rated according to their maximum DC power output (watts) under Standard Test Conditions (STC). Standard Test Conditions are defined by a module (cell) operating temperature of 25 °C , and incident solar irradiance level of 1000 W/m^2 and under Air Mass 1.5 spectral distribution. Since these conditions are not always typical of how PV modules and arrays operate in the field, actual performance is usually 85 to 90 percent of the STC rating.

Today's photovoltaic modules are extremely safe and reliable products, with minimal failure rates and projected service lifetimes of 20 to 30 years. Most major manufacturers offer warranties of twenty or more years for maintaining a high percentage of initial rated power output. When selecting PV modules, look for the product listing (UL), qualification testing and warranty information in the module manufacturer's specifications.

9.5 How a PV System Works

Simply put, PV systems are like any other electrical power generating systems, just the equipment used is different than that used for conventional electromechanical generating systems. However, the principles of operation and interfacing with other electrical systems remain the same, and are guided by a well-established body of electrical codes and standards.

Although a PV array produces power when exposed to sunlight, a number of other components are required to properly conduct, control, convert, distribute, and store the energy produced by the array.

Depending on the functional and operational requirements of the system, the specific components required, and may include major components such as a DC-AC power inverter, battery bank, system and battery controller, auxiliary energy sources and sometimes the specified electrical load (appliances). In addition, an assortment of balance of system (BOS) hardware, including wiring, overcurrent, surge protection and disconnect devices, and other power processing equipment. Fig 9.3 show a basic diagram of a photovoltaic system and the relationship of individual components.

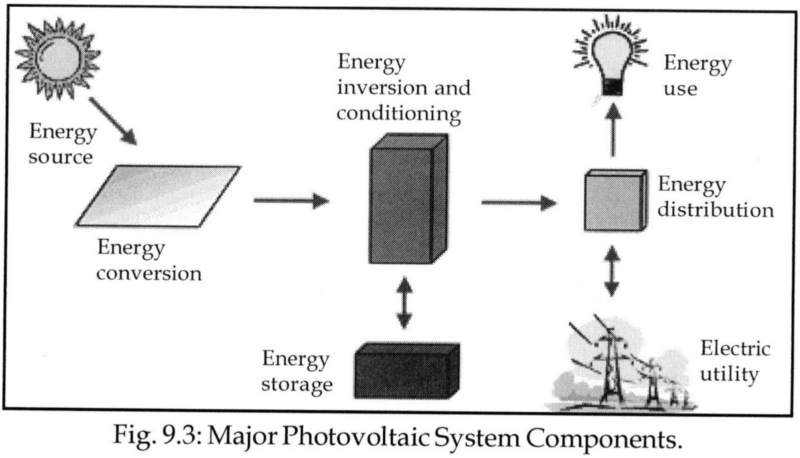

Fig. 9.3: Major Photovoltaic System Components.

9.6 Types of PV Systems

Photovoltaic power systems are generally classified according to their functional and operational requirements, their component configurations, and how the equipment is connected to other power sources and electrical loads. The two principle classifications are grid-connected or utility-interactive systems and stand-alone systems. Photovoltaic systems can be designed to provide DC and/or AC power service, can operate interconnected with or independent of the utility grid, and can be connected with other energy sources and energy storage systems.

9.6.1 Grid-connected (utility-interactive) PV systems

Grid-connected or utility-interactive PV systems are designed to operate in parallel with and interconnected with the electric utility grid. The primary component in grid-connected PV systems is the inverter, or power conditioning unit (PCU). The PCU converts the DC power produced by the PV array into AC power consistent with the voltage and power quality requirements of the utility grid, and automatically stops supplying power to the grid when the utility grid is not energized. A bi-directional interface is made between the PV system AC output circuits and the electric utility network, typically at an on-site distribution panel or service entrance. This allows the AC power produced by the PV system to either supply on-site electrical loads, or to back feed the grid when the PV system output is greater than the on-site load demand. At night and during other periods when the electrical loads are greater than the PV system output, the balance of power required by the loads is received from the electric utility This safety feature is required in all grid-connected PV systems, and ensures that the PV system will not continue to operate and feed back onto the utility grid when the grid is down for service or repair.

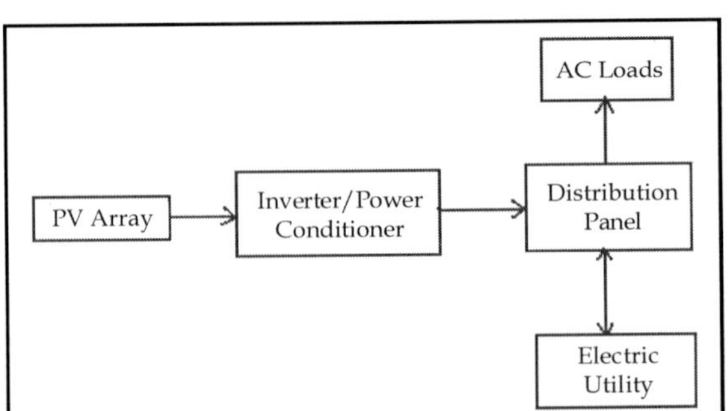

Fig.9.4. Diagram of grid-connected photovoltaic system.

9.6.2 *Stand-alone photovoltaic systems*

Stand-alone PV systems are designed to operate independent of the electric utility grid, and are generally designed and sized to supply certain DC and/or AC electrical loads. These types of systems may be powered by a PV array only, or may use wind, an engine-generator or utility power as an auxiliary power source in what is called a PV-hybrid system. The simplest type of stand-alone PV system is a direct-coupled system, where the DC output of a PV module or array is directly connected to a DC load (Fig.9.5). Since there is no electrical energy storage (batteries) in direct-coupled systems, the load only operates during sunlight hours, making these designs suitable for common applications such as ventilation fans, water pumps, and small circulation pumps for solar thermal water heating systems. Matching the impedance of the electrical load to the maximum power output of the PV array is a critical part of designing well-performing direct-coupled system. For certain loads such as positive-displacement water pumps, a type of electronic DC-DC converter, called a maximum power point tracker (MPPT) is used between the array and load to help better utilize the available array maximum power output.

PV Array → DC Load

Fig.9. 5: Direct-Coupled PV System.

In many stand-alone PV systems, batteries are used for energy storage. Fig 9.6 shows a diagram of a typical stand-alone PV system powering DC and AC loads. Fig 9.7 shows how a typical PV hybrid system might be configured.

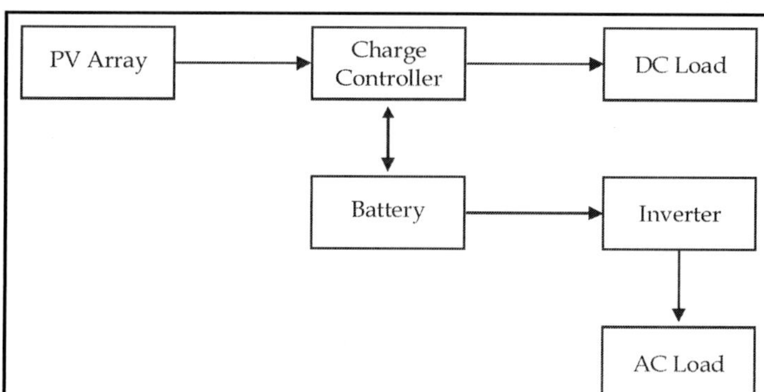

Fig 9.6. Diagram of Stand-Alone PV System with Battery Storage Powering DC and AC Loads.

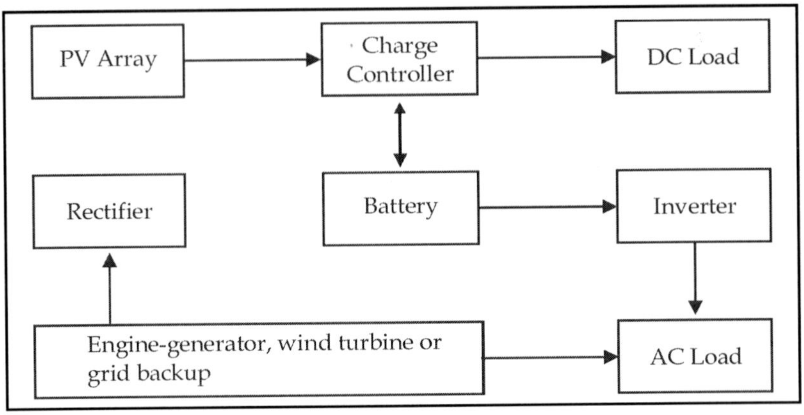

Fig 9.7. Diagram of photovoltaic hybrid system.

9.7 Materials for Photovoltaic Applications

There are various types of photovoltaic materials, which can be used to fabricate solar cells for converting sunlight into electricity. Few of these materials are:

(a) Homo junction solar cells

Solar cells made of from the same base material acros the p-njunction. Such as silicon solar cells.

(b) Hetro-junction solar cells

Solar cells having two base materials are known as hetro-junction solar cells. Such as SnO_2 on n-type silicon.

(c) Schottkg barrier Ms and MIS solar cells

In such a solar cells, the p-n junction is formed at a metal semiconductor (ms) interface. This can be prepared by depositing thin films of metals on base materials.

(d) Polycrystalline solar cells

The base materials of these cells are always crystalline Polycrystalline materials can be used for making solar cells using these film techniques.

(e) Amorphous solar cells

Materials like glasses are called 'Amorphous' materials, they have a short range order. The silicon can be grown in the amorphous state having photovoltaic properties. Such materials can be converted into p-n junction cells.

(f) Direct and indirect band gap solar cells

Materials like silicon are categorized under this class.

(g) Liquid interface solar cells

Such solar cells are formed by using top surface of a liquid electrolyte.

(h) Organic materials

Carbon based organic materials have also semi-conducting or photo conducting properties, those materials can also be used for making solar cells.

(i) Intermediate transition based solar cells

In these cells, front surface is coated by phosphors /fluorescent materials which may absorb high-energy radiation to avoid heat generation.

9.8 Efficiency of Solar Cells

Efficiency of a solar cell is defined as the ratio of energy output to the energy input from the sun. The energy output (watt -hour or W_h) indicates the amount of energy produced during the day. Most of the sun's energy reaching a solar cell is lost before it can be converted into useful electricity. The minimum amount of energy necessary to free an electron from its band varies with different semiconductor materials. Since solar cells are unable to respond to sunlight's entire spectrum, the solar cells cannot be 100 per cent efficient. Photon whose energy is less than that of the materials band gap, are not absorbed, resulting in a waste of about 25 per cent of the incoming energy. The energy content of photons above the band gap will be wasted surplus, re-emitted as heat or light .

Another factor that limits the cell efficiency is the inadvertent recombination of electrons and holes before they can contribute to an electric current. The natural resistance to electron flow in a cell also decreases cell efficiency. Such losses occur in three places; in the bulk of the base material, in the arrow tip surface and at the interface between the cell and the electric contacts leading to an external circuit.

Cell efficiency is also affected by temperature. Solar cells work best at low temperature as determined by their material properties. All cell materials give less efficiency as the operating temperature rises. One way to increase cell efficiency is to minimize the amount of light, which is reflected away from the cell's surface.

Thus, the efficiency of a solar cell is affected by several losses. Some of these losses are avoidable and other cannot be avoided under normal conditions of production and utilization. Some of these losses are given in Table 9.1.

There are few precautions required when working with the SPV system, which are listed below:

Table 9.1 Losses for Retarding Efficiency of Solar Cell

S. No.	Nature of loss	Amount of contribution in percentage
1.	Top surface constant losses	3
2.	Losses due to reflection at the top surface	1
3.	Photons not utilized due to their lower energy content than the band gap	23
4.	Excess energy of photons lost as heat	33
5.	Quantum efficiency losses if thickness of the cell is less than the minimum required thickness	0.4
6.	Collection losses; to minimize the collection loss better design is to be developed	Varies
7.	Voltage factor loss	20
8.	Curve factor loss	4
9.	Series resistance loss	0.3
10.	Shunt resistance loss	0.1
11.	Power left to be delivered	20 Approx.

(a) The best safety system is an alert mind, a skeptical nature and a slow hand.

(b) Never work on a PV system alone.

(c) Know the system before you start to work on it.

(d) Check your test equipment before going to the system site.

(e) Wear appropriate clothing and an approved electrical safety hat.

(f) Wear eye protection glasses particularly if working on batteries.

(g) Remove any jewellery.

(h) Wear dry leather gloves to reduce the probability of getting shock.

(i) Measure voltage from all conductors (on PV output circuit) to ground.

(j) Work with one hand whenever possible.

(k) Be skeptical, do not assume that configuration agrees with the electrical diagrams, that is current is not flowing in ground circuit, etc.

9.9 Applications of Solar Photovoltaic

The system run by the electricity generated by solar photovoltaic system can be classified for the following applications: telecommunication, railway signaling, cathodic protection, navigational aids, traffic warning light, remote instrumentation, crop spraying, water pumping, vaccine refrigeration, lighting system, battery charging, consumer products, educational kits, entertainment, offshore application, space applications, rural electrification, grid connected houses, hydrogen production and alarm systems, etc.

9.10 Present Status of Application of PV in India

Considerable experience has been generated in the country in the development and use of solar PV systems for several applications. The technology for the manufacture of solar cells, modules and systems has been developed indigenously. Systems for such applications as street lighting, domestic lighting, community lighting, water pumping for drinking water supply and irrigation, vaccine refrigeration, communication, etc. have been demonstrated and field- tested in the country. About 10,000 villages and hamlets have already been provided with amenities such as street lighting, lights in homes, community centers and night schools/adult education centers and community television, etc. About 1200 solar PV water-pumping systems have also been supplied to individual and institutional users as part of the SPV-demonstration programmes for meeting small irrigation and drinking water requirements. Some deep well water pumping systems have also been installed. About 80 solar PV centralized systems of capacity between 1 and 25 kw (peak) have been set up in different location in the country. The SPV systems have created a significant impact and are meeting the basic energy requirements of the people living in rural, remote and un electrified areas.

Currently there are 6 manufacturers of crystalline silicon solar cells and nearly 12 manufacturers of PV models including three public sectors companies among which Central Electronic Ltd. is the major production unit for Solar cell, PV modules and systems. Approximately 40 applications are being meet out through solar PV systems developed through CEL. The major areas where SPV systems are applicable are as follows:

1. Railway

The CEL has produced entire power requirement for signalling utilised at Railway stations in Northern Railways. They has also developed solar powered level crossing warning systems for Indian Railway.

2. Oil sectors

SPV based cathodic protection system for oil pipeline is also available. CEL has designed, developed, installed and commissioned all the cathodic protection system in western region of ONGC and low cost distributed cathodic protection system for Indian Oil Corporation.

3. Village electrification

Village electrification through solar PV system is very attractive option for the villages, which are not grid connected.

Apart from above specialized items, various commercialized gadgets are also available in market based on solar photovoltaic technology. These commercial products are given as below

(a) Solar Lantern

A typical solar lantern consists of a small photovoltaic module, a light source, a high frequency invertor/ballast, battery, charge controller and appropriate unit. During the day hours, the module-facing south is placed in the sun and it converts the solar radiation into electricity and charges the battery, which is connected, to the lantern through a cable. In the evening, the lantern with the charged battery is disconnected from the module and is available for indoor or outdoor use.

In another model, the module is integrated with the rest of the components in a flat house roof. In either case, there is no need for a mounting structure. However, a small and simple stand can be used to support the module. There is also a provision of central charging station in a village, where a sizable array are used to charge individual lanterns brought to the station daily by the villagers.

The solar lantern is a very simple unit, which can be used for lighting. It has no operating and maintenance cost except the change of battery after about four to five years of use (Fig. 9.8).

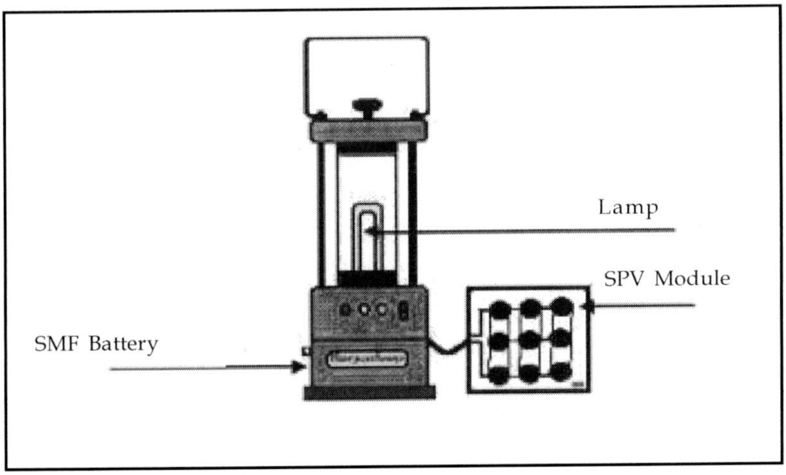

Fig. 9.8. Solar Lantern

(b) Street Lighting System (Fig. 9.9)

The solar PV based street lighting system has following three models, namely.

a) Single module, three hours operation, and fluorescent lamp with timer controls.

b) Two module, six hour operation, fluorescent lamp with timer control and

c) Two module, all night operation, with compact fluorescent lamp (CFL) and day/night sensor.

The governments have many schemes for the installation of the streetlights in rural areas and other remotely located places. As there is no installation cost or cost of setting up power lines such units are very cheap in far flung areas having no access to electricity.

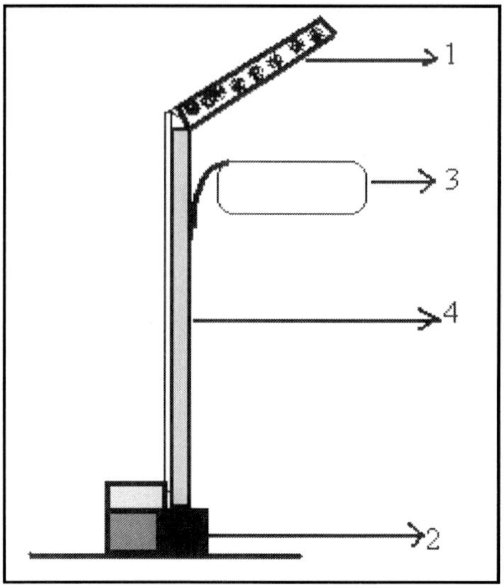

1. SPV module
2. Battery box
3. Lamp with charge controller
4. Lamp post

Fig. 9.9. Solar Street Lighting

(c) Domestic Lighting System

The following three types of domestic lighting systems are available in our country.

a) Fixed domestic lighting units (one module, 2 CFL and 7 W each with sealed battery and controls).

b) Portable lighting units-lanterns (10 W module, 5-7 watt CFL and sealed battery)

c) Service connections from photovoltaic power plants.

The each state government has set up a network for the purchase of the domestic units in rural and far-flung areas.

(d) Community PV Lighting System and PV Power Plants

For village level use, two types of systems are available:

a) A 300 Wp PV installation for the community centers in villages, largely comprising lights.

b) Village based PV power plants of 2 to 10 KW capacity or even higher capacity plants.

A typical village based PV power plant consists of an array, a battery bank housed in a small shed, a power conditioning unit, meters and controls and distributing system. The connected load may be for lighting, TV, radio and drinking water supply pumps. The cost of installation of such systems is shared by the state government, central government and villagers.

(e) Solar Water Pumping

Water pumping has a long history, so many methods have been developed to pump water with a minimum of effort. These have utilised a variety of power sources, namely human energy, animal power, hydro power, wind, solar and fossil fuels for small generators. The relative merits of these are laid out in Table 9.2 below.

Table 9.2: Comparison of Pumping Techniques

	Advantages	**Disadvantages**
Hand pumps	local manufacture is possible easy to maintain low capital cost no fuel costs	loss of human productivity often an inefficient use of boreholes only low flow rates are achievable
Animal driven pumps	more powerful than humans lower wages than human power dung may be used for cooking fuel	animals require feeding all year round often diverted to other activities at crucial irrigation periods
Hydraulic pumps (e.g. rams)	unattended operation easy to maintain low cost long life high reliability	require specific site conditions low output
Wind pumps	unattended operation easy maintenance long life suited to local manufacture no fuel requirements	water storage is required for low wind periods high system design and project planning needs not easy to install

□□□□□...

Table 9.2. Contd...

Solar PV	unattended operation low maintenance easy installation long life	high capital costs water storage is required for cloudy periods repairs often require skilled technicians
Diesel and gasoline pumps	quick and easy to install low capital costs widely used can be portable	fuel supplies erratic and expensive high maintenance costs short life expectancy noise and fume pollution

Solar pumps are used principally for three applications:

1. Village water supply
2. Livestock watering
3. Irrigation

A solar pump for village water supply is shown schematically in Fig 9.10.With village water supply a constant water demand throughout the year occurs, although there is need to store water for periods of low insolation (low solar radiation).

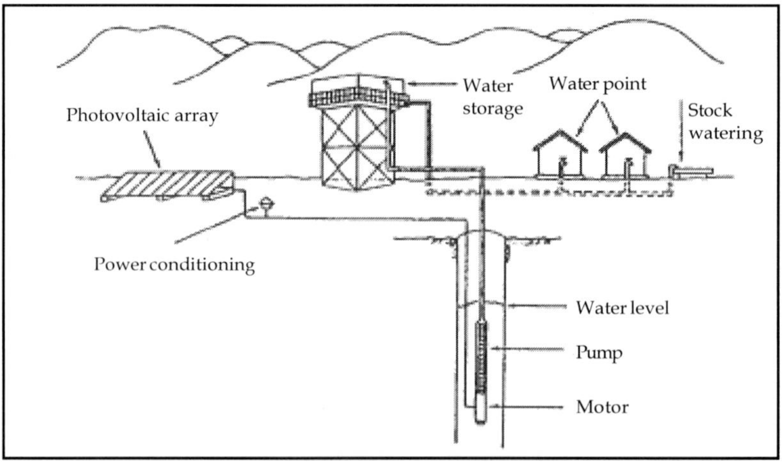

Fig. 9.10. Village Water Supply

A solar irrigation system (Fig 9.11) needs to take account of the fact that demand for irrigation water will vary throughout the year. Peak demand during the irrigation seasons is often more than twice the average demand. This means that solar pumps for irrigation are under-utilised for most of the year. Attention should be paid to the system of water distribution and application to the crops. The system should minimise water losses, without imposing significant additional head on the pumping system and be of low cost.

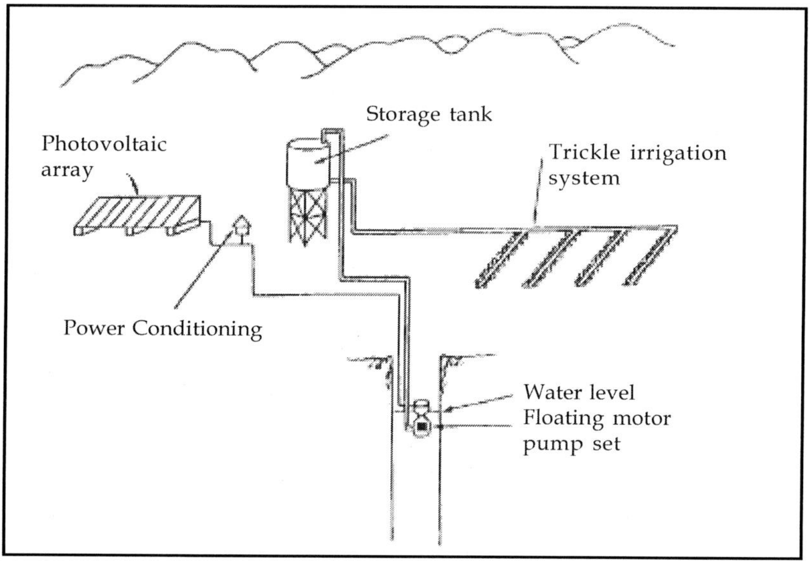

Fig 9.11. Solar Irrigation System

The suitability of major irrigation systems for use with solar pumps is shown in Table 9.3

Table 9.3 Suitability of Solar Pumps for Irregation

Distribution method	**Typical application**	**Typical head efficiency**	**Suitability for use With solar pumps**
Open Channels	50-60%	0.5-1m	Yes
Sprinkler	70%	10-20m	No
Trickle/drip	85%	1-2m	Yes
Flood	40-50%	0.5m	No

9.11 Design of Solar PV Technology for Water Pumping

(a) Daily Insolation levels

The power output from the PV array will depend upon the insolation and availability of sun per day. The insolation varies from one location to another, month to month because of seasonal and climatic changes. If water requirements are in the same range of the whole year, solar design calculations should be based on the month with the lowest insolation of the year. This will ensure adequate water supply through out the year. For irrigation water, the months with the lowest insolation often correspond to those in which crop demand for water is lowest. If water consumption varies round the year, the system design should be based on the ratio of water required to the insolation available. The month in which this ratio is largest will determine the PV array size.

(b) System sizing

The size of a PV water pumping system will depend on the water requirement, the total head and the solar insolation. The pattern of water use should also be considered in relation of system design and storage requirements enough to provide for daily water requirements and short periods of cloudy weather.

We can calculate the size of PV array by using the equation

$$E = (\rho\ g\ h\ v)\ /\ 3.6\text{x}10^6$$

where

E = Hydraulic energy required (kWh/day)

ρ = Density of water (1000 kg/m^3)

g = Gravitational acceleration (9.81 m/sec^2)

h = Total hydraulic head (m)

v = Volume of water required (m^3/day)

By putting these values the above equation reduces to

$$E = 0.002725\ hv\ \text{(kWh/day)}$$

The required PV array size can be calculated from equation

$$A = E\ /\ eFI$$

where

I = Average daily solar insolation (kWh/m^2)

F = Array mismatch factor (%)

e = Daily sub-system efficiency (%)

A = Cell / array area (m^2)

(c) Orientation and Direction of the PV Array

The PV array should be positioned in such a way that the sunlight is utilized to its maximum that is true south (in the Northern Hemisphere). The local declination which depends on the location and changes with the times should, however, be taken into account. When the direction north south has been found, the lines a and b are marked at right angles to the north-south line to indicate the position of the PV array. The PV array should not be shaded by obstructions like nearby building or trees.

(d) Determination of Tilt Angles

Module surface tilted at a right angle to the sun's rays catch the most sunshine per unit area. An angle equal to the local latitude is the closest approximation to that tilt or slope on annual basis. The tilt angle should be selected in accordance with the latitude in which the solar water pumping system is to be installed. If the water requirements do not remain uniform through out the year, a higher or lower PV array - tilt angle might be advantageous and lead to better system performance.

(e) Determining Peak Water Flow

The maximum required water flow in liters/hour would be approximately the system's requirement divided by the number of sun-hours. Dividing this figure by 360 second/hour gives the maximum expected water flow in liters/second. For example, to meet drinking water requirements of a typical village, 8000 liters of water requirements, 8 sun-hours per day of insolation is available, the peak flow rate from the system will be 8000 litres/5 sun hour = 1600 litres/ hour = 0.4 litres/second.

(f) Water Production

In designing a PV water pumping system, it is essential to know the amount of water a well can produce. A correctly operating pumping system should not exceed the well's water production. To determine the amount of water produced by a well, a portable pump is needed which is capable of pumping water at a rate at least as high as the peak-required rate.

First, measure the depth-to-water in the well. Install the pump and let it pump water until the water level in the well stabilizes. Now with the pump still operating, measure the depth-to-water again. Repeat the measurement at several time intervals to ensure that the water level has stabilized. Now measure the water flow rate by filling a container of known volume and measuring the time required to do so the flow rate should be as high as the peak flow rate required. If the water level drops to the bottom of the well, the well does not produce enough water for the village needs. So, either the well should be deepened or there has to be another well to meet the water requirements of the village.

(g) Wells

Wells are normally constructed to the depth of the local water level plus and additional depth to account for pumping draw down. Well must be designed to provide uncontaminated water free from bacteria and abrasive that wear out pump components. Usually, wells are lined with RCC, plastic or steel wall casing extending to the aquifer.

Well yield is a crucial design parameter that must be considered relative to the maximum water-pumping rate. Water extraction rate should not exceed the well yield or the water table may fall below the suction of the pump. Many pumping system failures are directly attributable to well design and yield problems.

(h) System Configuration

Water pumping systems can be configured for three general application ranges: These are shallow water table, low-flow intermediate and deep water table application ranges of commercially available motor pump sets.

(i) Shallow well application (7 meters water level). The photovoltaic water pumping system for shallow well applications will consist of a well with a surface mounted single-stage centrifugal pump directly coupled to a DC motor.

(ii) Intermediate/Deep well Application (20-40 meters water depth).

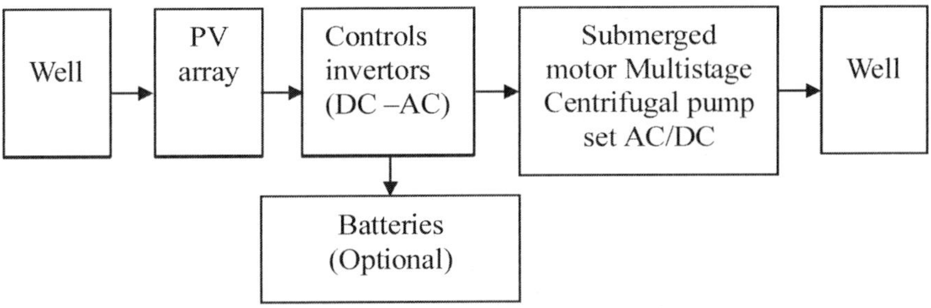

Low flow (30,000 liters/day or less)

The surface-mounted centrifugal pumps are suited for locations where the water levels do not exceed 7 meters. Beyond 7 meters, the centrifugal pump will lose suction and will not be able to lift water from the well.

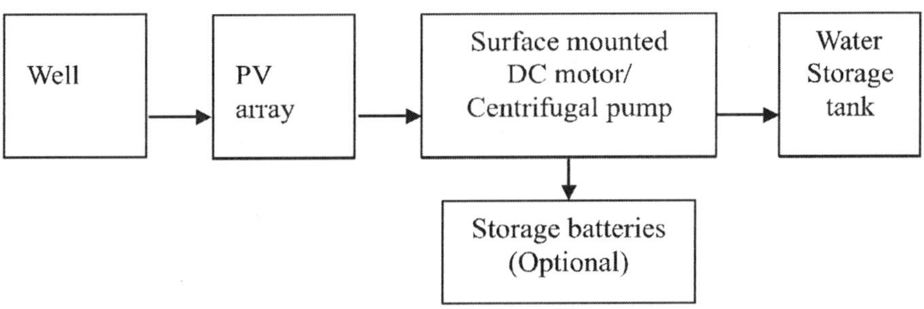

For intermediate and deep well applications, the choice is a jack pump or submersible motor pump set. At water depths of 20 meters, jack pumps are good choice for water if demand is below 30,000 liters per day. In this, mounted DC motor drives a jack pump. The system comprises power-conditioning equipment to match the cyclic continuous photovoltaic power output.

High flow (over 30,000 liters per day)

To meet water requirements of more than 30,000 liters per day, a submersible motor pump set is recommended. The system will consist of a well, PV array, controls and invertors and AC motor pump set in case of an AC system. A DC motor pump set is used to PV array capacity of 1 kWp. No invertors are required for a DC system. This type of system uses a multi-stage centrifugal pump with a submersible motor. For peak power demands beyond 1 kWp and AC system energizing AC motor, which operates, multistage centrifugal pump is assumed.

(i) Operation and maintenance

Operation and maintenance requirements for a solar photovoltaic water pumping system are minimal. Maintenance is required only of the motor/pump set such as the lubrication of parts and replacement of brushes.

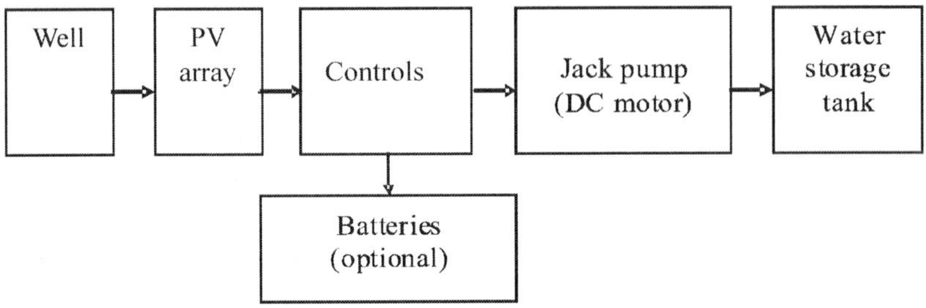

The cleaning of the top glass surface of the photovoltaic modules with a duster is required regularly, in some cases once a week or once in 15 days. Cleaning of the glass surfaces with water may be done, if required. Batteries can be considered only in the case of motors requiring high starting current. If the batteries are of stationary, tubular type not much maintenance is needed and only addition of distilled water in required.

10

Energy from Biomass

10.1 General

The term "biomass" generally refers to renewable organic matter generated by plants through photosynthesis. Materials having organic combustible matter is also referred under biomass.

Biomass is an important fuel source in our overall energy scenario. Biomass is produced through chemical storage of solar energy in plants and other organic matter as a result of photosynthesis. During this process conversion of solar energy in sugar and starch, which are energy rich compounds takes place. The chemical reaction of photosynthesis can be written as:

$$6CO_2 + 6H_2O + \text{sun light} \rightarrow C_6\,H_{12}\,O_6 + 6O_2 + 636 \text{ kcal.}$$

It indicates that the storage of 636 kcal is associated with the transfer of 72 gm carbon into organic matter. Biomass can be directly utilized as fuel or can be converted through different routes into useful forms of fuel. In fact, biomass is a source of five useful agents, which start with '5F' like food, fodder, fuel, fiber and fertilizer. Further biomass has many advantages like.

(1) It is widely available

(2) Its technology for production and conversion is well understood.

(3) It is suitable for small or large applications

(4) Its production and utilisation requires only low light intensity and low temperature (5-35°C)

(5) It incorporates advantage of storage and transportation

(6) Comparatively, it is associated with low or negligible pollution.

10.2 Biomass Classification

Biomass includes plantation that produces energy crops, natural vegetable growth and organic wastes and residues. The biomass classification is illustrated in Fig. 10.1. It can be grouped as:

(1) *Agricultural & Forestry Residues:* Silviculture crops.

(2) *Herbaceous Crops:* Weeds, Napier grass.

(3) *Aquatic and marine biomass:* Algae, water hyacinth, aquatic weeds, plants, sea grass beds, kelp and coral reep etc.

(4) *Wastes:* Municipal solid waste, municipal sewage sludge, animal waste and industrial waste etc.

India produces about 532.44 million tonnes of agricultural residues every year. Similarly, 273 million cattle population produces on an average about 433 million tonnes of dung annually. Fuel wood is another major source of biomass in India. The fuel wood consumption in India is estimated to be about 227 million tonnes per year. Some of biomass sources are given below:

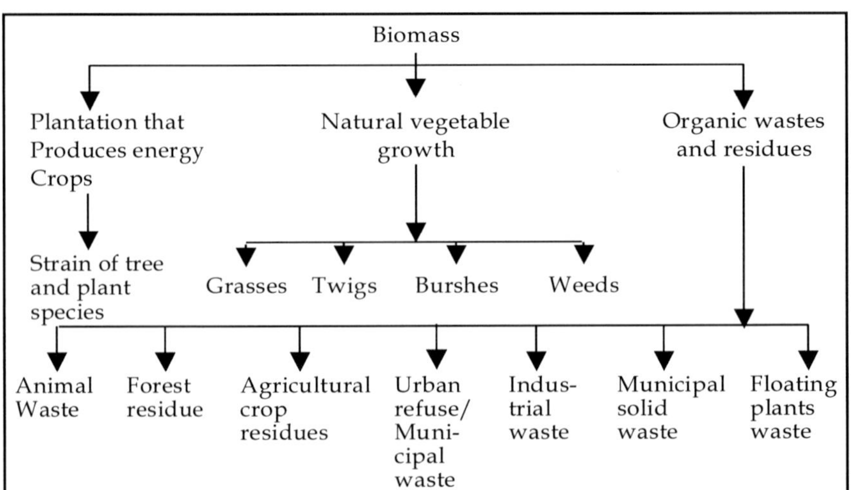

Fig. 10.1 Classification of Biomass

10.2.1 *Energy plantation*

This term refers to an area that is used to grow biomass for energy purposes. The idea behind energy plantation programme is to grow selected strains of tree and plant species on a short rotation system on waste or arable land. The sources of energy plantation depend on the availability of land and water and careful management of the plants. As far as suitability of land for energy plantation is concerned following criterion is used-

(1) It should have minimum of 60 cm annual precipitation

(2) Any arable land having slope equal to or less than 30% is suitable for energy plantation.

The economic of energy plantation depends on the cost of planting and availability of market for fuel. Whereas these two factors are location specific and these varies from place to place. Further productivity of this programme depends on the microclimate of the locality, the choice of the species, the planting spacing, the inputs available and the age of harvest. There are many suitable species for energy plantation, few of them are :

(1) *Acacia nilotica* (Babul),

(2) *Acacia auriculiformis* (Bengali Babul),

(3) *Dalbargia sissoo* (Shisham),

(4) *Eucalyptus comaldulusis* (Eucalyptus),

(5) *Leucaena lencocephala* (Subabul),

(6) *Prosopis chilensis* (Perdesi Babul),

(7) *Prosopis Juliflora* (Vilayati Babul),

(8) *Tamarix articulate* (Jungle jalebi),

(9) *Tamarindus indica* (Imali),

(10) *Albizza lebbek* (Siris).

10.2.2 *Animal waste*

The animal waste is also an organic material, which have combustible property. The amount of wastes an animal can produce depends on the size of the animal, feed and the condition of confinement. The animal waste is a rich source of fuel. The dung cakes so prepared with animal wastes can be used for meeting cooking energy requirement in rural and semi-urban areas. Further, recycling of animal waste anaerobically can produce a combustible gas known as biogas. Thus, animal waste is a essential ingredient for fuel production through cakes and anaerobic digestion. The anaerobic digestion is a process through which not only combustible gas is obtained, but also enriched manure is produced which can help in replacement of chemical fertilizer. The production of dung from an animal is generally assumed to be directly proportional to the live weight of the animal. Average dropping from each animal species is depended on the physical characteristics of animal. It is given in Table 10.1

Dead weight of animal is also a waste, which through full decomposition acts as enriched manure for increasing productivity of the land.

Table 10.1 Average Dropping by Animal

S. No.	Animal	Normal average dropping per animal per day (kg)
1.	Cow	10.0
2.	Buffalo	15.0
3.	Bullock	12.5
4.	Camel	6.0
5.	Horse	10.0
6.	Sheep	1.0
7.	Pig	2.25
8.	Chicken (2 kg weight approx)	0.18

10.2.3 Forest waste / residues

A variety of residues are produced as a result of the growth and harvesting of trees from the forest. It include sawdust, bark generated in sawmills, tree removed in intermediate cuttings, trees and shrubs removed from the under story in evenaged stands and dead trees as a result of natural causes, including disease, insect attack, fire and storms. The availability of forest residues, which depend on the production of fuel wood, can be estimated on the basis of tree or plant species, forest area, and type of soil and availability of water.

A conventional tree plantation, without water and fertilizer, produces small amounts of wood and other residues. Further, various forms of management techniques such as fertilization, pest control, irrigation, genetic plant selection and thinning can improve productivity considerably. Short rotation hardwood can yield upto 4 tons per acre a year. The higher the productivity, the less land area needed to produce a given amount of woody biomass.

Wood and bark residues are generated in mills, and also in forests along with same foliage residues. Logging residues also represent a widely and readily available source of forest residues. These generally consist of the above ground portions of trees not removed in such logging operations. The most important characteristics of forest residues are their calorific values, ultimate analysis, moisture content, particle size and bulk density. The soft wood species have calorific value between 4510 and 5335 kcal per kg, where as hard wood species have calorific value between 4345 and 4840 kcal/kg. Bark generally has higher calorific value than wood, ranging from 4554 to 5643 kcal/kg for soft wood species and from 3888 to 5219 kcal/kg for hard wood species. The calorific value for softwood foliage has been estimated in the range of 4399 to 4977 kcal/kg.

10.2.4 Agricultural crop residues

Agricultural crop residues are considered one of the chief sources of biomass for energy production, because these are available in abundance as natural resources during the production of agricultural produce. Further, these permit easy collection and procurement and long term storage. It has been observed that annually in our country about 200 million tonnes of agricultural residue such as sugarcane leaves, rice husk, cotton stalks etc. is available.

The amount of agricultural residue depends on the extent of agricultural crop production. Therefore, the quantity of crop residues would also depend upon the type of crop sown. High productivity of agricultural crops can be accomplished by increasing area, changing the ratio of the biomass of reproductive to that of vegetation organs and by other morphophysiological properties. The availability of principal agro wastes in India is given in Table 10.2

Table 10.2 Estimated Production of Agricultural Residues.

S.No.	Products	Annual Production
1.	Rice husk	11.8 Million tons
2.	Wheat straw	58.1 Million tons
3.	Wheat husk	13.9 Million tons
4.	Groundnut shells	1.8 Million tons
5.	Groundnut straw	8.6 Million tons
6.	Corn cobs	11.2 Million tons
7.	Corn stalk	12.2 Million tons
8.	Corn straw	6.2 Million tons
9.	Cotton stalk	18.5 Million tons
10.	Sugarcane leaves	5.0 tons / ha
11.	Bermuda grass	15.0 tons /ha
12.	Congress grass	6.0 tons/ha

10.2.5 Urban refuse/municipal wastes

The amount of waste produced in urban areas depends upon the population and standard of living. The quantity of waste produced daily by a person is estimated at 2.27 kg. The composition of urban and municipal waste varies depending upon the season and also from one region to another. The municipal waste consists of solid wastes as human excreta (night soil), household wastes (garbage), city wastes (public utilities and sanitation services) and commercial wastes (institutional activity). It also includes liquid form such as domestic sewage and effluent from community institutional activities. The availability of urban and municipal wastes can be assessed according to the population and waste generated per person per day. At present the urban refuse is disposed of in landfills and also by burning in municipal incineration. However, these operations are not desirable on account of their contribution to environmental pollution and because the available space for land fills is scarce. Therefore, the conversion of urban refuse into most useful from of energy may solve the problem of disposal. The energy sources represented by this refuse are renewable.

Presently disposal of this waste acquire significant attention for environment protection and as well as for energy production. There are numbers of ways for these wastes can be recycling and resources recovered in terms of fuel gas, manure and liquid fuel etc.

10.2.6 Agro-industrial wastes

Recently, because of rapid industrialization, the amount available as a waste created a need for its safe disposal on account of its environmental degradation properties. These industrial wastes can also be used as a bio-energy source.

Industrial waste consists of the materials discarded from industrial operations. These include wastes obtained from product processing, packaging, shipping office and other wastes. Since the composition of wastes varies from one industry to another, they will exhibit different heating values. These wastes may be solid waste as well as liquid waste generated because of industrial and production activities.

Both municipal and industrial wastes possess relatively high heating values, which allows them to be incinerated with very little, additional fuel. Incineration is a solution that can reduce the amount of refuse for disposal to about one-tenth of its original volume. The heat generated by incineration of refuse can be used for steam production which can also be used for power generation or heating purposes (industrial process heat).

10.2.7 Floating plants or marine biomass

At present, floating water plants (e.g. water hyacinths) are pest plants in many rivers, lakes and ponds in tropical and semitropical areas in the world. The rate of growth of water hyacinths is rapid and in nutrient-rich ponds net productivity of up to 25 tons of dry product per acre per year has been reported. The floating water plants can be used as a fuel for methane production and also as a cattle feed. However, in view of the land water constraints, large-scale plantation of such plants may have little relevance unless these are grown in sewage or soil water. The production of marine biomass as an energy source has several attractive considerations, the most important being the large area of ocean available for such purposes. However, the disadvantages include the meager state of existing knowledge and the difficulty of operating under oceanic conditions.

10.3 Biomass Production Technique

The biomass production should be viewed with number of points starting from preparation of soil to planting seedlings and upto harvesting of the biomass. In fact, these all operations are combined for biomass production to achieve multiple benefits. Thus, a careful planning is required for biomass production, which consists of integration of different techniques and improved methods of production. The general sequence for biomass production is illustrated in Fig. 10.2. It includes step starting from site survey, nursery techniques, transplanting techniques and maintenance of the plantation.

10.3.1 *Site survey*

As far as site survey for biomass production is concerned, it should watch with the species of tree and shrubs best suited to the area. Many information are needed for proper site survey, such as climate, soil, topography, vegetation, biotic factors, water table levels, availability of supplementary water sources and distance from nursery etc.

10.3.2 *Planting site selection*

It is based on the information of the site survey. The best site is one, which on plantation will lead to the establishment of a successful plantation.

However, the choice of the planting site is limited to lands, which are not suited for agriculture or livestock production. The boundaries of the planing site should be marked with boundary posts after the area has been chosen. When there is a danger of trespassing and damage by grazing animals a boundary fence should be established. Of course, fencing is costly and therefore it should only be built when other means of protection are not effective. Once the forest plantation is well established with sufficiently tall trees, the fences can be removed and reused at another planting site. When roads and other passageways traverse the planting site, they also should be contained with fences.

10.3.3 *Species selection*

Once the selection of site plantation is over, then following points should be consider in the selection of the tree and species.

1. A species selected for a particular site, should suit to the site and should remain healthy throughout the anticipated rotation and should exhibit acceptable growth and yield.

2. For a successful planting, performance data can have to be extrapolated from one locality to another. Results from a locality where a tree or shrub species is growing (either naturally or as an exotic) strictly apply only to that locality.
3. When reliable information shows a close similarity between the site to be planted and that on which the species already is successful, this particular site is recognized for large-scale planting.
4. The selection of tree and shrub species through the use of analogous climates is important as a first step in selection. This step is amplified by an evaluation of localized factors (for example soil, slope and biotic factors).
5. The ability to match closely a planting site and a natural habitat may not preclude the need for species trials, since climatological or ecological metering may not reveal the adaptability of a species.

10.3.4 Preparation of the planting site

The preparation of the planting site is more important in the biomass establishment programme.

The site preparation is meant for removing competing vegetation from the site, creating conditions that will enable the soil to catch and absorb as much rainfall as possible and providing good conditions for the planting, including a sufficient volume of routable soil. The site preparation also includes the creation of conditions where the dangers from fire and pests are minimized. The site preparation is directed toward giving the seedlings a good start with rapid early growth. It reduces surface runoff for increasing the moisture in the soil and at the same time it helps in elimination of hard pans of the soil.

10.3.5 Preparation of the soil mixture

The soil mixture for the containers as well as for seedbeds should have following characteristics.

(1) It should be relatively light and cohesive
(2) It should have a good water retention capacity
(3) It should possess a high organic and mineral content
(4) It should contain adequate nutrients, which usually are supplied in the form of artificial fertilizers.

(5) The quantity of soil needed in a containerized nursery operation is directly related to the size of the containers used. For example, to fill 1,00,000 small containers, 200 cubic meters of soil are needed, where as 442 cubic meters of soil are needed to fill 1,00,000 of the largest containers (16 times more).

(6) The Soil used in nursery work should be sufficiently acidic in nature. For most tree and shrub species, the pH should not be higher than 6.0.

Damping off is a common and serious disease in many forest nurseries. It can occur either in seedbeds or in containers after transplanting. This disease is caused various fungi. A watery - appearing construction of the stem at the ground line generally is visible evidence of the disease. Damping off is favored by high humidity, damp soil surface, heavy soil, cloudy weather, an excess of shade, a devise stand of seedlings and alkaline conditions.

One of the best preventive measures for this disease is to maintain dry soil surface through cultivation, to reduce the sowing density and to thin the seedling to create better aeration at the ground line.

10.3.6 Sowing of seed

Sowing of seed should match to the ideal sowing time. In order to determine the ideal sowing time, one counts backwards from the beginning of the planting season to identify the number of months required to raise high quality planting stock. The time to raise high quality planting stack depends on number of factors like; type of tree or shrub species, climatic conditions and nursery conditions.

It is essential for each planting project to prepare its own sowing schedules for the locally important tree and shrub species. With large amount of seeds, not all should be sown at the same time. It is better to spread out the sowing dates with one or two week intervals.

A common method of raising seedlings has been to sow the seed in seedbeds or seedling trays, and then to transplant into containers as soon as the plants are sufficiently large to handle. The seedbeds can be either broadcast sown or sown in lines. Small seed should be mixed with some kind of inert fine material to facilitate even distribution.

10.3.7 Method of sowing

Seedling trays or boxes commonly used for the production of seedlings, which subsequently are transplanted into containers. The trays can be easily moved to the beds for the transplanting operation. A good tray can be made from kerosene or diesel tins by cutting them lengthwise into halves. The bottom of the tray should be perforated to allow drainage or irrigation by absorption. In order to assist drainage, the bottom of the tray can be filled with a 2 cm layer of gravel or charcoal.

Once the seeds are sown in seedlings or seedling trays, they should be covered with sand, which should be pressed down gently to establish a good contact between seeds and soil. Depth of sowing depends on the size of the seed. The normal depth of sowing is from 1 to 3 times the diameter of the seed, but in many instances, it can be desirable to sow slightly deeper to avoid the washing of seeds from the soil by irrigation or heavy rains.

10.3.8 Transplanting of seedling into containers

A plant that is grown either in seedbed or seedling trays, in which it was originally sown and than it has to lifted for final replacement is technically known as a "seedlings". When a seedling is lifted and replanted in the nursery in another bed or container, it is therefore, termed a "transplant". The transplanting of seedlings, also referred to as "pricking" which is done primarily to induce of better development of the root system by increasing the number of fine absorbing root lets.

Transplanting should be done before the seedling has acquired a large, heavy root system, but after it has developed a strong stem. Normally, this stage occurs after the complete unfolding of the cotyledons and during the unfolding of the first leaves. Transplanting is a delicate and time-consuming operation and requires precise organization and sufficient labour. Therefore, there is a tendency to sow directly into the containers when possible, which has the advantages saving of time, labor, reducing the losses and retardation in plant growth caused by transplanting.

10.3.9 Transport of seedlings to the planting site

Generally, plants are damaged during transport to the planting site. Therefore, adequate care must be taken to avoid mishandling of plants during loading and unloading from vehicles. Even, plants require protection during transportation, as the airflow can cause breakage of stems or drying. Therefore, it is also important that the containers as packed tightly, so that they cannot move. When possible, plants be transported in the planting season on cool, cloudy or even rainy day to prevent desiccation during transport. Following are few suggestions, which should be kept in mind during transportation of seedlings to the planting site.

1. Always prefer the packing of container- raised plants for transportation.
2. Put packed container- raised plants in trays the load into vehicles.
3. The tins, which have been used for seedling trays can be used for transporting container plants.
4. Shipping schedule should be planned to avoid delays and to allow proper disposition of the plants immediately upon arrival.
5. Normally, plants should arrive one day ahead of planting.
6. Where shape and watering facilities are available, supplies can be brought in advance of several days.
7. As soon as the plants arrive at the planting site, they must be watered and if necessary, heeled in a cool, moist, shaded place until they are needed for planting.

10.3.10 Maintenance of the plantations

Once a plantation has been established, than it is necessary to protect the plantation against damaging. A variety of cultural treatments also can be required to meet the desired purpose of the plantation-protection of plantation against weather effect, fire damage, pests and insects, wild animals is essential. The cultural treatments include weeding, thinning, pruning and maintaining desired spacing between trees.

10.4 Harvesting of Biomass

Biomass harvesting is decided when trees and shrubs attain the "optimum size" for the wood production. From a biological point of view, trees and shrubs should not be cut until they have at least grown to the maximum size required for product utilisation. In general, the average annual growth of trees and shrubs increases slowly during the initial year of establishment, reaches a maximum and then falls more gradually in subsequent years. Hence, trees and shrubs usually should not be allowed to grow beyond the point of maximum average annual growth, which is the age of maximum productivity. As far as harvesting of biomass is concerned following factors be considered.

1. *Biological factors:* Age of maximum productivity, average growth rate.
2. *Pathological factors :* Growth in terms of mortality and the amount of defect in living trees.
3. *Entomological factors :* Forest composition, age, structure and vigour.
4. *Silvicultural factors :* Seed production, characteristics, methods of obtaining regeneration, competition from less desirable tree species and maintenance of desirable soil conditions.
5. Local harvesting techniques.
6. Available man power.
7. Existing market outlets.

Biomass harvesting should be well planned and organized in order to make the best use of the raw material while keeping labour input and production cost low and minimising damage to the environment. A variety of different harvesting systems can be applied. The factors on which harvesting systems depend are as follows-

1. Species of wood
2. Size and assortment (fuel wood, poles or logs)
3. Type of forest (man made or natural)
4. Type of cut (thinning or clear cut)
5. Kind of regeneration (artificial, coppice and natural)
6. Terrain (flat, steep and swampy)
7. Accessibility (roads, waterways) and
8. Means of transport (manual, animal and motorized)

In all these cases, good planning and organization of work depends on following factors-

1. The assessment of the volume to be harvested.
2. The determination of the assortments to be produced.

3. The determination of wood storing places, skidding lines and felling direction.
4. Clearly - instructed and skilled supervision and workers.
5. Availability of the necessary hand tools and maintenance and other equipment.
6. Clear separation of working areas for individual work teams and different operations (filling and transport).

10.5 Harvesting Methods

These are several different harvesting methods that allow the plant to re-generate through sprouting. These are as follows-

(a) Coppicing

It is one of the most widely used harvesting methods in which the tree is cut at the base, usually between 15-75 cm above the ground level. New shoot developed from the stamp or root. These shoots are some times referred to as sucker or sprouts. Management of sprouts should be carried out according to use. For fuel wood the number of sprouts to container to grow, should depend on desired sizes of fuelwood. If many sprouts will be allowed to grow for a long period, the weights of the sprouts will became heavy and the sprouts may tear away from the main trunk. Several rotation of coppicing are usually possible with many species. The length of the rotation period depends on the required tree products from the plantation. It is suitable method for production of fuelwood. Most eucalyptus species and many species for leguminous family, most of naturally accessing shrubs can be harvested by coppicing (Fig. 10.2a).

(b) Pollarding

It is the harvesting systems, in which the branches including the top of the tree are cut, at a height of about 2 meter above the ground and the main trunk is allowed standing. The new shoots sprout or emerged from the main stem to develop a new crown. This results into continuous increase in the diameter of main stem although not in height. Finally when the tree losses its sprouting vigour, the main stem is also cut for use as large diameter poles. An advantage of this method over coppicing is that the new shoots are high enough off the ground that they are out of reach of most grazing animals. The neem tree

(*Azadirachata indica*) is usually harvested in this manner. The branches may be used for poles and fuel wood (Fig. 10.2b).

(c) Lopping

In this method most of the branches of the tree are cut. The fresh foliage starts sprouting from the bottom to top of the denuded stem in spite of severe defoliation, surprisingly quickly. The crown also regrows and after a few years, the tree is lopped again. The lopped trunk continues to grow and increase in height, unless this is deliberately prevented by pruning it at the top (Fig. 10.2c).

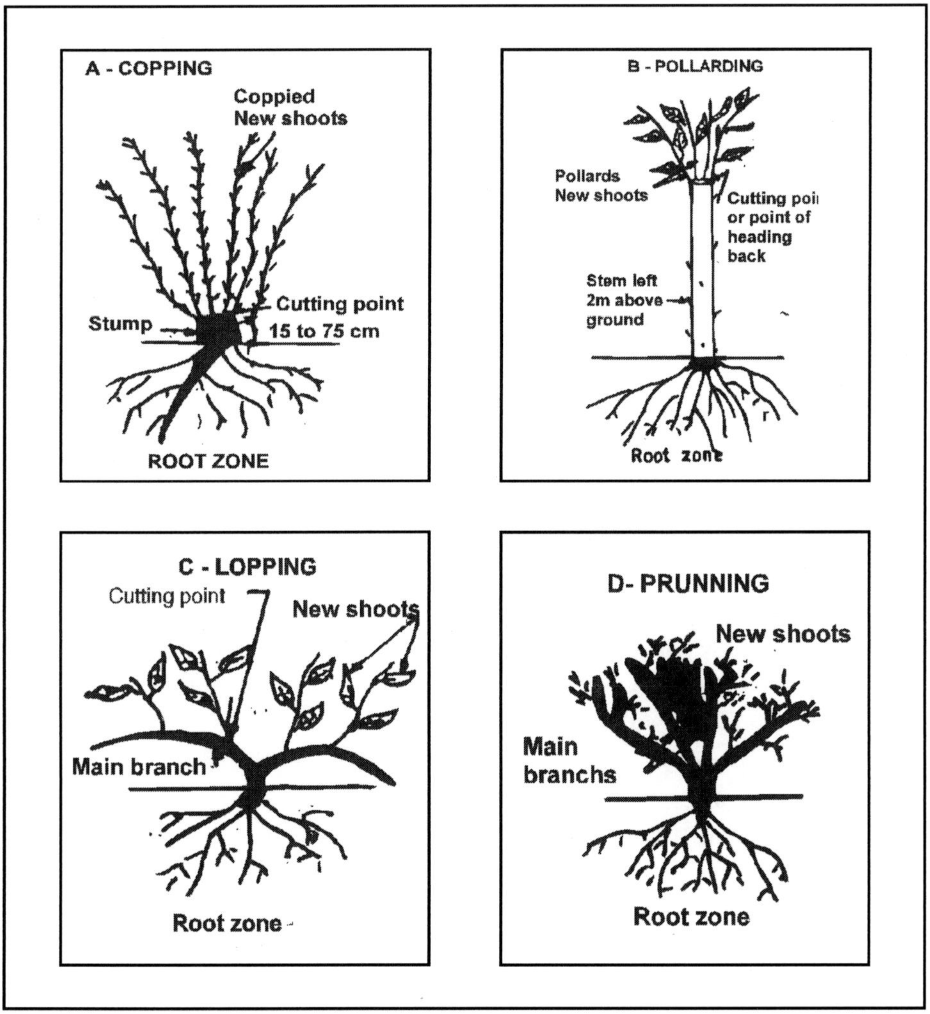

Fig. 10.2 Different Harvesting Methods

(d) Pruning

It is very common harvesting method. It involves the cutting of smaller branches and stems. The clipped materials constitute a major source of biomass for fuel and other purposes, such as fodder mulching between tree sows. It is also often required for the maintenance for fruits and forages trees, alley cropping and lives fences. The process of pruning also increases the business of trees and shrubs for bio fencing. Root pruning at a distance of 2-3 meter from the hole is effective to reduce border tree competition for water and nutrient with the crops (Fig. 10.2d).

(e) Thinning

It is a traditional forestry practice and in fuelwood plantation, it can also be of importance. The primary objectives of the thinning are to enhance diametric growth of some specific trees through early removal of poor and diseased tree to improve the plantation by reducing the competition for light and nutrients. Depending on initial plant density, initial thinning can be used for fuelwood or pole production. If maximum biomass production is the main objectives of the plantation, regardless of quality, thinning may not be needed.

10.6. Biomass Processing for Fuel Use

Biomass can be processed into a number of ways for fuel use. The biomass can be used as an energy source material for heating, mechanical work or for electricity generation. It can be combusted directly to produce heat or it can be converted into a secondary fuel or into an energy carrier, such as steam, compressed air, electricity etc. Secondary conversion is through physical, chemical or biological processes or a combination of these.

10.6.1 Physical processes

Physical processes involves many unit operations such as drying (change in the moisture constant), size reduction (change in surface area to volume ratio) and densification (change in density).

(a) Drying

It is method of removing moisture content of biomass. The heat generation of biomass is dependent on the moisture content, therefore, drying of biomass is essential for enhancing its calorific value. It has been observed that about 9 per cent of energy value of biomass are lost in reducing the moisture content from 30 to 9 per cent, but if it is not reduced, the decrease in calorific value is about 26 per cent.

(b) Size Reduction

The size reduction of biomass is done to convert it to a convenient transportable, storable and usable form. These processes include tree cutting (removing stems and branches from tree), log cutting to domestic size or convenient size for use and log cutting to small billet form for use.

(c) Densification

As the name indicates, this process is to increase bulk density of biomass for efficient and convenient transportation and handling. It also reduces requirement of bulk storage space. The physical dimensions and the combustion characteristics of the fuel become homogeneous and uniforms because of the required particle size, porosity and density and as a result of it efficient energy conversion is established.

Densification processes are basically of three types.

(i) *Binderless densifcation technique* : Where no external binders are added for compacting the materials. However, it is done at very high temperature (200°C) and at high pressure (1500 kg/cm^2).

(ii) *Binder* : Densification technique- Where pressure requirement is less through same types of binding materials such as tar, molasses, sodium bentonite, resins or wax.

(iii) *Pyrolysis* : Densification technique - In this technique, firstly the biomass is carbonized and the charcoal obtained is powdered and then compacted in required shape.

Depending upon final shape of product obtained through densification process, the process is terms as briquetting, pelletization and cubing.

(d) Biomass Briquetting

Bulk of the energy in the rural areas for cooking and heating purposes is derived from firewood, agro and forest residues and to some extent from cow dung. Use of firewood leads to deforestation and inefficient combustion of cow dung, agro and forest residues leads to pollution problems. In addition, materials like pine needles are major causes of forest fires.

To solve these problems and to save wood, the agro and forest residues can be upgraded to convenient and smokeless briquetted fuels. These materials can be converted into small sized briquettes with holes (beehive briquettes) through partial pyrolysis process. These briquettes can burn with a smokeless and clean flame, ideal for domestic and small scale heating purposes. The agro and forest residues can also be directly briquetted by compaction at high pressures. These briquettes are suitable for large scale and industrial applications. The agro and forest residues in briquetted fuel form has added advantage of easy transportation to the place of work.

10.7 Type of Gasifiers

Design of gasifier depends upon type of fuel used, power rating required and whether gasifier is portable or stationary. Gasifier are classified according to how the air blast is introduced in the fuel column. History of gasification reveals serveral designs of gasifiers. The most commonly built gasifiers are classified as :

1. Updraft gasifier
2. Downdraft gasifier
3. Twin-fire gasifier
4. Crossdraft gasifier
5. Fluidized bed gasifier
6. Other gasifier

10.7.1 Updraft gasifier

An updraft gasifier has clearly defined zones for partial combustion, reduction, and pyrolysis. Air is introduced at the bottom and act as counter current to fuel flow (Fig. 10.3). The gas is drawn at higher location. The updraft gasifier achieves the highest efficiency as the hot gas passes through fuel bed and leaves the gasifier at low temperature. The sensible heat given by gas is used to preheat and dry fuel. Disadvantages of updraft gasifier are excessive amount of tar in raw gas and poor loading capability. Hence it is not suitable for running vehicle.

10.7.2 Downdraft gasifier

In the updraft gasifier, gas leaves the gasifier with high tar vapour which may seriously interfere in the operation of internal combustion engine. This problem is minimized in downdraft gasifier. In this type, air is introduced into downward flowing packed bed or solid fuels and gas is drawn off at the bottom (Fig-10.4). A lower overall efficiency

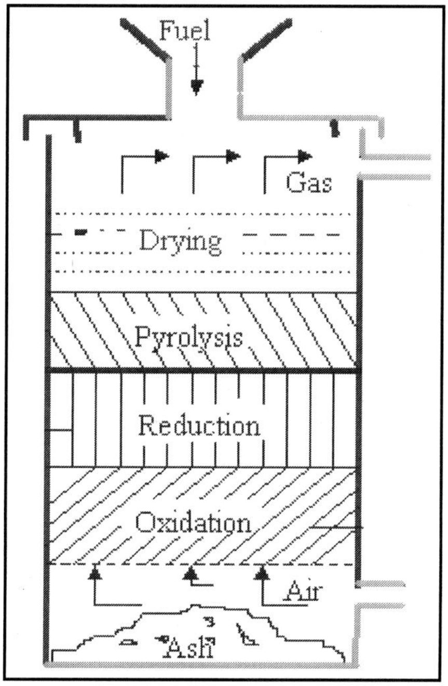

Fig.10.3 Updraft Gasifier

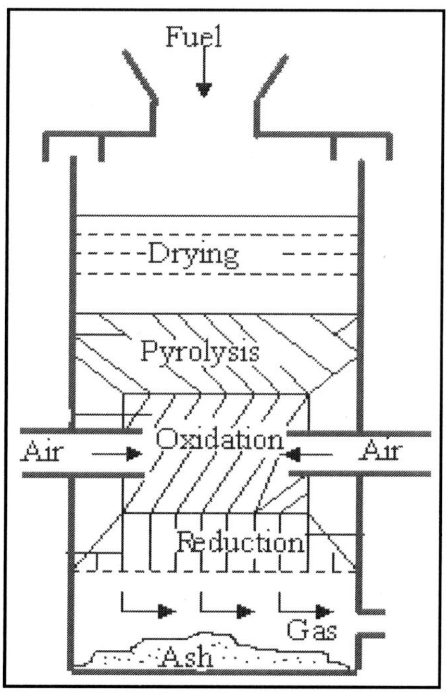

Fig. 10.4 Downdraft Gasifier

and difficulties in handling higher moisture and ash content are common problems in small downdraft gasifiers. The time (20-30 minutes) needed to ignite and bring plant to working temperature with good gas quality is shorter than updraft gas producer. This gasifier is preferred to updraft gasifier for internal combustion engines

10.7.3 Twin-fire gasifiers

The advantage of co-current and counter-current gasifiers are combined in a so called twin-fire gasifier. It consists of two defined reaction zones. Drying, low-temperature carbonisation, and cracking of gases occur in the upper zone, while permanent gasification of charcoal takes in lower zone. The gas temperature lies between 460 to 520°C. Total process takes place in under pressure of -30 mbar. Twin-fire gasifier produces fairly clean gas (Fig. 10.5).

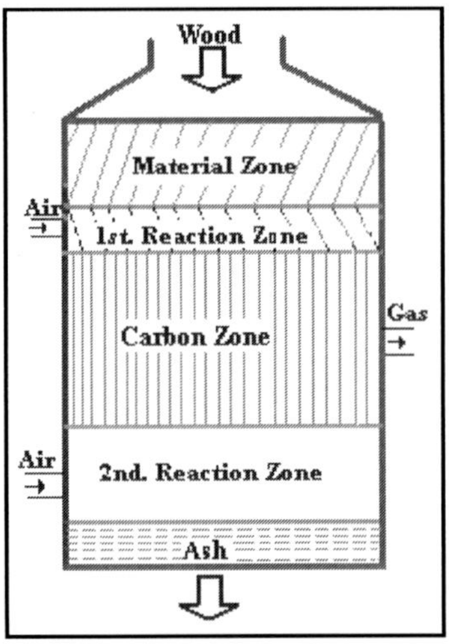

Fig. 10.5 Twin-Fire Gasifier

10.7.4 Crossdraft gasifier

Crossdraft gasifier, although they have certain advantages over updraft and downdraft gasifiers. However, they are not of ideal type. The disadvantages such as high exit gas temperature, poor CO_2 reduction and high gas velocity are the consequence of the design. Unlike downdraft and updraft gasifiers, the ash bin, fire and reduction zone in crossdraft gasifiers are separated. This design characteristics limit the type of fuel for operation to low ash fuels such as wood, charcoal and coke. The load carrying capacity of crossdraft gasifier is quite good due to concentrated partial zones which operates at temperatures up to 2000°C. Start up time (5-10 minutes) is much faster than that of downdraft and updraft units. The relatively higher temperature in cross draft gas producer has an obvious effect on gas composition such as high carbon monoxide, and low hydrogen and methane content when dry fuel such as charcoal is used. Crossdraft gasifier operates well on dry air blast and dry fuel (Fig. 10.6).

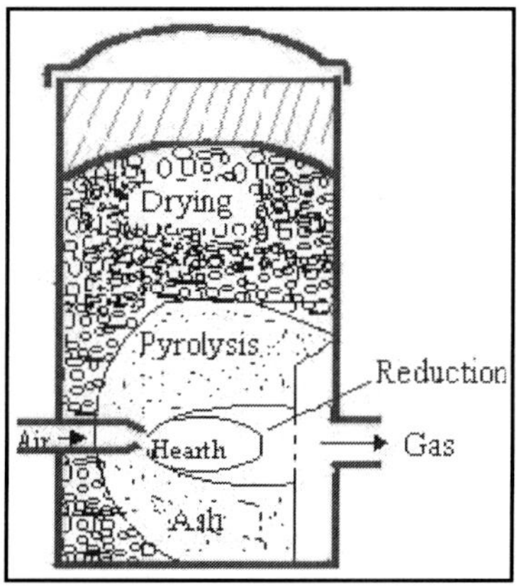

Fig. 10.6 Crossdraft Gasifier

10.7.5 Fluidized bed gasifier

The operation of both up and down draught gasifiers is influenced by the morphological, physical and chemical properties of the fuel. Problems commonly encountered are: lack of bunker flow, slagging and extreme pressure drop over the gasifier

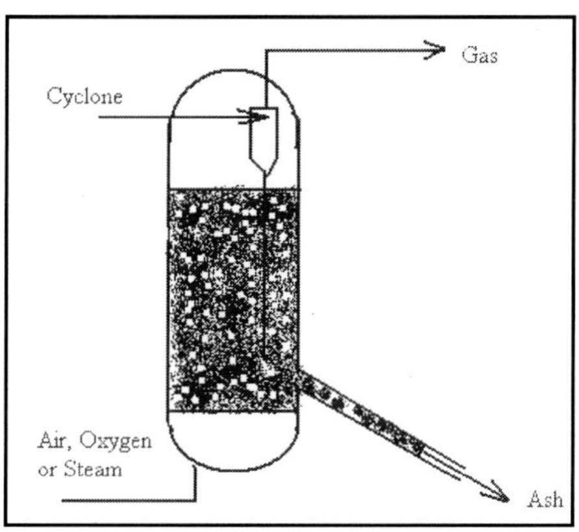

Fig. 10.7 Fluidized Bed Gasifier

A design approach aiming at the removal of the above difficulties is the fluidized bed gasifier illustrated schematically in Fig.10.7.

Air is blown through a bed of solid particles at a sufficient velocity to keep these in a state of suspension. The bed is originally externally heated and the feedstock is introduced as soon as a sufficiently high temperature is reached. The fuel particles are introduced at the bottom of the reactor, very quickly mixed with the bed material and almost instantaneously heated up to the bed temperature. As a result of this treatment the fuel is pyrolysed very fast, resulting in a component mix with a relatively large amount of gaseous materials. Further gasification and tar-conversion reactions occur in the gas phase. Most systems are equipped with an internal cyclone in order to minimize char blow-out as much as possible. Ash particles are also carried over the top of the reactor and have to be removed from the gas stream if the gas is used in engine applications.

10.7.6 Other type of gasifiers

Although updraft, downdraft or crossdraft gasifier have been the most commonly used systems, there is a wide variety of gasifiers which do not really fit into any of these categories and are classified as other gasifier. Some units are built to combine the advantages of crossdraft with downdraft or updraft gas producers.

10.8 Producer Gas and It's Constituents

Producer gas is the mixture of combustible and non combustible gases. The quantity of gases constituents of producer gas depends upon the type of fuel and operating condition.

The producer gas delivered from the gasifier has the following average chemical composition.

1. Healiny value : 4.5 to 6 MJ/m^3
2. CO - 15 to 30% by volume basis. Although carbon monoxide posses higher octane number of 106, its igniton speed is low. This gas is toxic in nature. Hence, human operator need to be careful while handling gas.
3. H_2 - 18-22%, octane number 60-66 and it increases the ignition ability of producer gas. CH_4 & H_2 are responsible for higher heating value of producer gas.
4. CH_4 upto 4%
5. CO_2 - 5 to 15%.

6. N_2 - 45 to 60%.
7. Water vapour in the producer gas occur due to moisture content of air introduced during oxidation process, injection of steam in gasifier or moisture content of biomass fuels.

10.9 Advantages of Gasification

Conversion of solid biomass into combustible gas offers all the advantages associated with using gaseous and liquid fuels such as clean combustion, compact burning equipment, high thermal efficiency and a good degree of control. In locations where biomass is already available at reasonable low prices (e.g. rice mills, coffee/corn processing units, sugar mills, etc.) or in applications utilizing fuelwood (e.g. institutional cooking, silk reeling units, etc), gasifier systems offer definite economic advantages. These are at present made especially attractive with the incentives offered by the Ministry of non-conventional energy sources, Govt. of India.

10.10 Application of the Technology

(a) Steam generation

Small-sized baby boiler is used at present in many small industries like food processing industries. Boilers can be retrofitted easily to burn producer gas. Installation of gasifier helps in replacing wood/ biomass burning boilers which are generally bulky and polluting. Users of LDO and furnace oils can shift to firewood /wood waste briquettes at specific sites where biomass is available at low cost.

(b) Institutional/large-scale cooking

Fuelwood is presently used in large quantities for cooking in hostels, hospitals, hotels, marriage parties and sweet shops. Thermal efficiency of large stoves using firewood is low (approximately 10 per cent) and necessitates the use of large quantities of firewood. Use of gasifier reduces fuelwood consumption by about 50 per cent. Power delivered can be varied over a large range making cooking faster Depending on the availability of biomass and space, LPG users can shift to firewood/ biomass briquettes.

(c) Silk reeling industry

Large cottage basin units for silk reeling in south India use large quantities of firewood. Thermal efficiencies of cottage basin ovens are very low (10–14 per cent). Use of gasifiers can reduce firewood consumption by 50 per cent. Clean flue gas obtained by burning producer gas can be used for drying of pupae (silkworm). Producer gas can also be used for stifling of cocoons to kill pupae.

(d) Crematoriums

Currently, about 400 kg of fuelwood is required for every cremation. Annually about 2.44 million tonnes of fuelwood is consumed. Pollution of rivers like the Ganga and the Yamuna along the banks of which dead bodies are burnt. Air pollution due to large quantities of biomass usage. Water pollution due to throwing of under-burnt bodies into the rivers. Gasifiers can reduce fuel consumption drastically. Gasifiers would be sentimentally acceptable alternatives for cremation, as the body is burnt in flames. Gasifiers can alleviate pollution problems to a large extent.

(e) Drying

Present methods of drying cash crops like cardamom, tea, and areca nut using firewood are very inefficient. Usage of gasifiers can reduce fuel consumption by at least 50 per cent. Flue gases obtained by burning producer gas can be used either directly or indirectly (by employing a compact plate-fin heat exchanger) for drying or curing of various items including fruits, vegetable, fish, and other food products.

(f) Power generation

Producer gas can be cleaned to a high degree (approximately 50 mg/Nm^3 tar/dust with simple equipment. Existing diesel engine sets can easily be converted to operate on dual (diesel + producer gas) fuel mode. Diesel replacement possible upto 80%, even with large variation of load.

11

Biodiesel-Energy Options

11.1 Biodiesel Chemistry

Biodiesel is a mixture of methyl esters of long chain fatty acids like lauric, palmitic, stearic, oleic and so on. It is produced by the transesterification of animal fats and vegetable oils – all of which belong to a group of organic esters called triglycerides. Typical examples are rape/canola oil, soyabean oil, sunflower oil, palm oil and its derivatives, etc. from vegetable sources, beef and sheep tallow and poultry oil from animal sources and also from used cooking oil. The chemistry is basically the same irrespective of the feedstock. The chemistry of biodiesel is given in Fig. 11.1.

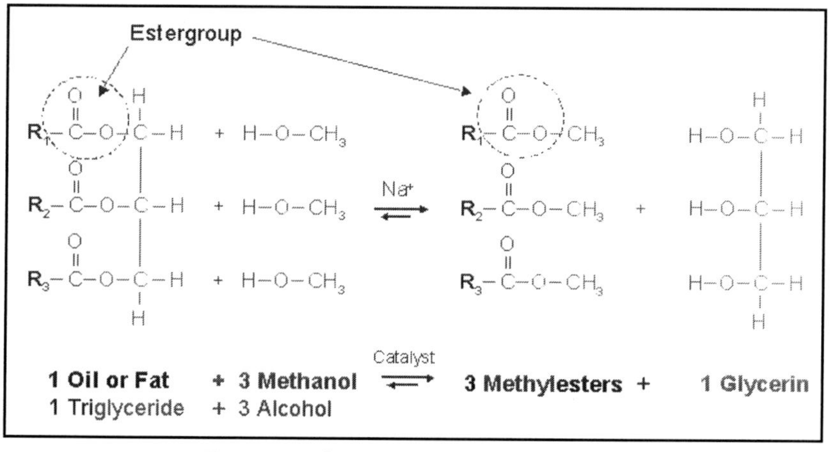

Fig 11.1. The Chemistry of Biodiesel

The abbreviation R_1, R_2 and R_3 are symbolic representations of the fatty acid chains, which can vary in molecular chain length from typically C8 to C22, and also their degree of unsaturation. An example is Oleic acid, which has 18 carbon atoms and one double bond:

CH_3-CH_2-CH_2-CH_2-CH_2-CH_2-CH_2-CH_2-CH=CH-CH_2-CH_2-CH_2-CH_2-CH_2-CH_2-CH_2-COOH

or more simply: CH_3-$(CH_2)_7$CH=CH-$(CH_2)_7$COOH

The abbreviation C(18:1) is also often used.

11.2 Biodiesel Process

The production of biodiesel involves intensively mixing of methanol with oil or fat in the presence of a suitable catalyst and then allowing the lighter methyl ester phase to separate by gravity from the heavier glycerol phase. However, as with most organic reactions the degree of conversion depends on the equilibrium reached as well as the influence of other reactions. Achieving product quality is also very important. An example of process configuration designed to optimize the production of biodiesel is illustrated in Fig.11.2

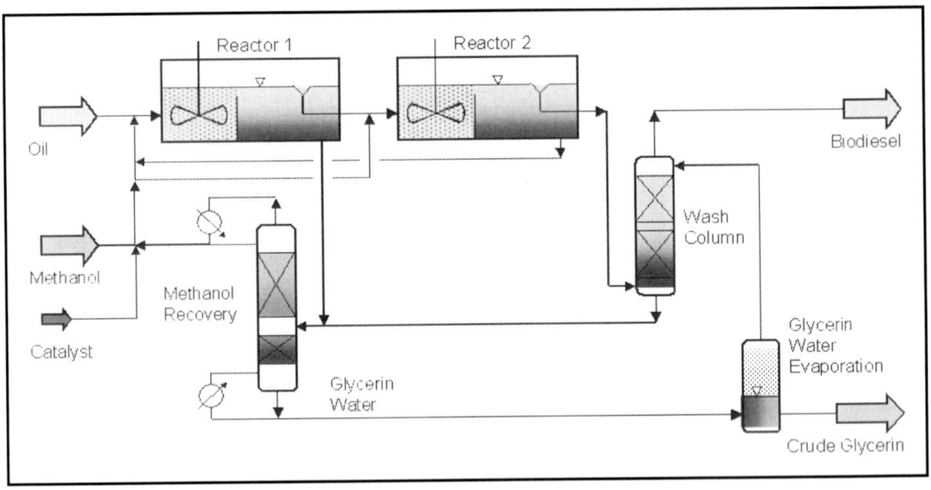

Fig 11.2. Biodiesel Process

Oil, methanol and sodium methylate catalyst are mixed in reactor 1 and allowed to separate into two phases. The lighter methyl ester/ oil phase is then mixed with additional methanol and catalyst in reactor 2 and again allowed to separate. This second reactor stage forces the transesterification process in the direction of the methyl ester and hence maximizes the biodiesel yield and quality. The lighter phase is then washed with water to remove any residual glycerol or methanol dissolved in the ester phase, followed by a vacuum drying to yield biodiesel.

The denser glycerol phase from reactor 2 contains excess methanol and catalyst, so this is recycled to the front end of reactor 1. The denser glycerol phase leaving reactor 1 will still contain unreacted excess methanol which is recovered by distillation in the Methanol Recovery Column and recycled to the front of reactor 1.

The wash water from the water wash column is also fed to the methanol recovery column for methanol recovery. Thus all of the methanol is consumed in the production of methyl ester. The bottom of the Methanol Recovery Column is processed in the Glycerin Water Evaporation Column to recover water for recycle to the Wash Column, leaving a Crude Glycerin byproduct containing 80-85 per cent glycerin.

The key features of biodiesel process are:

- Technology applicable to multiple feedstocks
- Continuous process at atmospheric pressure and nominally 60°C
- Dual Reactor System operating with a patented Glycerin Cross Flow configuration for maximized conversion
- Reaction using excess methanol, but with full methanol recycle to avoid any losses
- Closed loop water wash recycle to minimize waste water generation
- Clear phase separation by special gravity process (no centrifuges necessary)
- Based on suitable feedstock, this produces biodiesel to current world standards
- Raw Glycerin to BS 2621.

The crude glycerin can be further upgraded to pharmaceutical glycerin standard EU Pharmacopoeia 99.5 by distillation, bleaching if required, and vacuum drying.

11.3 Key Issues in the Manufacturing of Biodiesel

i. *Quality of Feedstock:* The performance of the process and the quality of the biodiesel produced during transesterification are dictated by the quality of feedstock. Therefore, it is important to feed an oil to the process that has been pretreated:

- To remove water as this causes dissociation of the catalyst
- To remove free fatty acids as they will produce soaps which interfere with phase separation and distillation
- To remove insoluble material, sediments and carbon residues as they can fail the corresponding biodiesel specification tests
- To remove phosphatides as they interfere with phase separation and they generate slime and gumming characteristics

ii. *Nature of Feedstock:* Nature of feedstock is also one of the important characteristics for production of biodiesel. Not all vegetable oils can be converted to biodiesel. Certain methyl esters, particularly those from saturated fatty acids, display high melting temperatures, and even when blended with many other methyl esters, this characteristic may lead to failing certain biodiesel specification tests. All diesel fuel is filtered before being injected into a diesel engine. If the biodiesel or the biodiesel blend is sufficiently cold then some of these particular methyl esters will crystallize out of solution and cause the filter to plug up, thus starving the diesel engine. The particular test is termed as the "Cold Filter Plug Point" and the European Standard EN 14214 has different classes depending on climate and the time of year - Class A for warm climates the CCPP is 5°C (max) but for very cold climates class F has a CFPP of –20°C (max).

Similar Tests such as "Cloud Point" and "Pour Point" also provide a means of differentiating between different biodiesels. For example the Pour Point of Rape Methyl Ester (RME) is –15°C, Soy Methyl Ester (SME) is –3°C and Tallow Methyl Ester (TME) +16°C. Obviously TME will fail the test first, but there are several ways of overcoming this problem, such as limiting the amount of biodiesel added to say 2% (B2), adding a Pour Point Suppressant or fractionating the biodiesel product to reduce the level of the responsible methyl esters.

iii. *Influence of Feedstock on Glycerin Quality:* Glycerin is used extensively throughout the pharmaceutical and cosmetic industries. Particular emphasis is placed on sourcing the glycerin from non-animal feedstocks, such that it can be termed kosher quality. Such glycerin commands significantly higher prices than that sourced from tallow or other animal sources. It is even becoming more important to avoid potentially contaminating vegetable based glycerin with any other material. With the coming growth in biodiesel production pressure will be put on glycerin prices and finding markets for non-kosher glycerin The sale of glycerin is an important revenue stream for the economics of biodiesel production.

11.4 Characteristics of Biodiesel

Biodiesel as automotive fuel has similar properties to petrodiesel and as such can be directly used in existing diesel engines with no or minor modifications. It can be used alone or mixed in any ratio with petrodiesel. The most common blend is B20, a mix of 20 per cent biodiesel with 80 per cent petroleum diesel. Biodiesel has 11 per cent oxygen by weight and essentially contains no sulphur or aromatics.

Physical Properties

Some of the physical characteristics of biodiesel are given in Table 11.1.

Table 11.1. Physical Properties of Biodiesel

S.No.	Properties	Values
1.	Specific gravity	0.88
2.	Viscosity @ 20 C (centistokes)	7.5
3.	Cetane Index	49
4.	Cold filter Plugging Point (°C)	-12
5.	Net Heating Value (Kilojoules/Liter)	33,300

11.5 Biodiesel Specifications

The key components, which determine the quality of biodiesel are monoalyl esters, dialkyl esters, residual vegetable oil, free glycerin, reactant alcohol, free fatty acids and the residual catalyst. In December 2001, American Society of Testing & Materials (ASTM) issued a specification (D6751) for biodiesel (B100) which is presented in Table-11.2. A provisional specification for B20 biodiesel is also notified by the ASTM known as ASTM PS121 (Table-11.3). Table-11.4 shows a comparison of selected properties of biodiesel and petrodiesel.

Table 11.2. ASTM Specification (D6751) for B100

Property	ASTM Method	Limits	Units
Flash Point	D93	130 min.	°C
Water & Sediment	D2709	0.050 max.	% Volume
Kinematic Viscosity (40 C)	D445	1.9-6.0	mm^2/sec
Sulfated Ash	D874	0.020 max.	% mass
Sulphur	D5453	0.05 max.	% mass
Copper Strip Corrosion	D130	No.3 max.	
Cetane	D613	47 min.	
Cloud Point	D2500	Report	°C
Carbon Residue (100% Sample)	D4530*	0.050 max.	% mass
Acid Number	D664	0.80 max.	Mg KOH/gm
Free Glycerin	D6584	0.020 max.	% mass
Total Glycerin	D6584	0.240 max.	% mass
Phosphorous Content	D4951	0.001 max.	% mass
Distillation Temperature (90% Recovered)	D1160	360 max.	°C

*The carbon residue shall be run out on the 100% sample.

Table 11.3 ASTM PS121 for B20

Property	ASTM Method	Limits	Units
Flash Point	D93	100 min.	°C
Water & Sediment	D2709	0.050 max.	% Volume
Kinematic Viscosity (40 C)	D445	1.9-6.0	mm^2/sec
Sulfated Ash	D874	0.020 max.	% mass
Sulphur	D5453	0.0015 max.	% mass
Copper Strip Corrosion	D130	No. 3 max.	
Cetane Number	D613	46 min.	

Contd...

Table 11.3 Contd...

Cloud Point	D2500	Report	°C
Carbon Residue (100% Sample)	D4530	0.050 max.	% mass
Carbon Residue (Ramsbottom)	D524	0.090 max.	% mass
Acid Number	D664	0.80 max.	Mg KOH/gm
Free Glycerin	D6584	0.020 max.	% mass
Total Glycerin	D6584	0.240 max.	% mass

Table 11.4. Selected Fuel Properties for Petrodiesel & Biodiesel

Fuel Property	**Petrodiesel**	**Biodiesel**
Fuel Standard	ASTM D975	ASTM PS121
Fuel Composition	C10-C21 HC	C12-C22 FAME
Lower Heating Value, Btu/gal	131,295	117,093
Kin. Viscosity, @40 C	1.3-4.1	1.9-6.0
Specific Gravity, kg/l @ 60 F	0.85	0.88
Density, lb/gal @ 15 C	7.079	7.328
Water, ppm by wt.	161	0.05% max.
Carbon, wt%	87	77
Hydrogen, wt%	13	12
Oxygen by dif. Wt%	0	11
Sulphur, wt%	0.05 max.	0.00-0.0024
Boiling Point, °C	188-343	182-338
Flash Point, °C	60-80	100-170
Cloud Point, °C	-15 to 5	-3 to 12
Pour Point, °C	-35 to -15	-15 to 10
Cetane Number	40-55	48-65
Stoichometric Air/Fuel Ratio	15	13.8
BOCLE Scuff, gm	3,600	>7,000
HFRR, microns	685	314

11.6 Toxicity of Biodiesel

Impacts of biodiesel on human health is also a significant criteria for including suitability of the fuel for commercial applications. Health effects can be measured in terms of fuel toxicity to the human body as well as health impacts due to exhaust emissions. Tests conducted by the Wil Research Laboratories investigated the acute oral toxicity of pure biodiesel fuel as well as B20 in a single dose study on rats, which concluded that biodiesel is not a toxic and there is no hazards anticipated from ingestion incidental to industrial exposure. The acute oral LD50 (lethal dose) is greater than 17.4-g/kg-body weight, which by comparison is far safer then even table salt. According to National Institute for Occupational Safety & Human Health, a 96-hr. lethal concentration of biodiesel for bluegills was greater than 1000 mg/l and this aquatic toxicity is deemed as insignificant. Other related effects of biodiesel are given below:

- Very mild human skin irritation. It is less than the irritation produced by 4 per cent soap and water solution.
- It is biodegradable. It degrades at least 4 to 5 times faster then conventional diesel fuel.
- Biodiesel has a flash point of about 300 °F well above conventional diesel fuel.
- Spills of biodiesel can decolorize any painted surface if left for long.
- There is no tendency for the mutagenicity of exhaust gas to increase for a vehicle running on biodiesel (20 per cent RSME+ 80 per cent diesel).

11.7 Prospective Feedstock in India

Oil can be extracted from a variety of plants and oilseeds. Under Indian condition only such plant sources can be considered for biodiesel production which are not edible oil in appreciable quantity and which can be grown on large-scale on wastelands. Moreover, some plants and seeds in India have tremendous medicinal value, considering these plants for biodiesel production may not be a viable and wise option. Considering all the above options, probable biodiesel yielding trees in India are:

(1) *Jatropha curcas* or Ratanjot

(2) *Pongamia pinnata* or Karanj

(3) *Calophyllum inophyllum* or Nagchampa

(4) *Hevea brasiliensis* or Rubber seeds

(5) *Calotropis gigantia* or Ark

(6) *Euphorbia tirucalli* or Sher; and

(7) *Boswellia ovalifololata.*

Among all the above prospective plant for as biodiesel production, Jatropha curcas stands at the top. One hectare Jatropha plantation with 4400 plants under rain fed conditions can yield about 1500 literes of oil. It is estimated that about 3 million hectares plantation is required to produce oil for 10 per cent replacement of petrodiesel. The residue oil cake after extraction of oil from Jatropha can be used as organic fertilizers. It is also estimated that one acre of Jatropha plantation could produce oil sufficient to meet the energy requirement of a family of 5 members and the oil cake left out when used as fertilizer could cater to one acre.

11.8 Economic Feasibility

The expected cost of extracted oil in Indian condition would be Rs. 20-25 per litre. However, the price of commercially available bio-diesel is presently Rs. 33 - 38 per litre. With recent increase in oil prices, it is essential to look for substitutes of fossil fuels for both economic and environmental benefits to the country. The bio-fuel extraction will disseminate technology and provide better energy services at the village level.

11.9 Environmental Benefits

There are number of environmental benefit through use of biodiesel, which are given as below:

- The use of bio-fuel avoids fossil fuel use and hence avoids CO_2/ CO emission in atmosphere
- It is a promising alternative fuels source for future
- Substantial reduction of unburned hydrocarbons, carbon monoxide and particulate matter
- Decrease the solid carbon fraction of particulate matter
- Increase in the green cover as result of plantations would check soil erosion and retain moisture and soil nutrients.

- Positive ecological benefits in terms of lending support to biodiversity, especially since degraded lands are involved

Biofuels also help the environment because the plants grown to make these fuels take greenhouse gases such as carbon dioxide out of the air and fix it in their roots, stems and leaves. Much of this carbon dioxide gets sequestered in the soil, reducing the overall level of carbon dioxide in the atmosphere.

12

Biogas Technology

12.1 Introduction

Nature has a provision for destroying and disposing off wastes and dead plants and animals. This decay or decomposition is carried out by tiny micro-organisms called bacteria. Making of farm-yard manure (FYM) and compost is also through decomposition of organic matter (OM). When a heap of vegetable or animal waste and weeds etc., die or decompose at the bottom of backwater or shallow lagoons, bubbles can be noticed rising to the surface of water. Sometimes these bubbles burn with dancing flame at dusk. This phenomenon has puzzled man for a long time. It was only during the past hundred years that Scientists unlocked this secret as the decomposition process. The gas thus produced was and is still called "Marsh Gas." The technology of harnessing this gas under artificially created conditions is known as Biogas Technology.

12.2 Biogas Technology

Biogas Technology has a very significant role to play in integrated agricultural operations, rural sanitation, large-scale dairy farms & sewage disposal etc. It is estimated that cattle dung, when passed through a Biogas unit, yields 30-40 per cent more net energy and about 35-45 per cent more Nitrogen in manure as compared with that obtained by burning dung cakes and ordinarily prepared compost, respectively. Besides, from a biogas plant both the products are obtained. There are about 250 million bovine (cattle and buffalo) population in India and one biogas unit for small family requires about 3-5 cattle heads, thus about 10 million family size plants fed on cattle and buffalo dung can be installed.

12.2.1 Process description

Biogas generation is a process widely occurring in nature and can be described as a biological process in which biomass or organic matter, in the absence of Oxygen is converted into Methane and Carbon dioxide. It is characterized by low nutrient requirement and high degree of waste stabilization process where biogas is one of the two useful products; the other being enriched organic manure in the form of digested slurry. It is essentially a three stage process involving following reactions:

(1) Hydrolysis (2) Acid formation (3) Methane generation.

For all practical purposes the first two steps are often defined as a single stage, i.e. hydrolysis and acid formation stages are grouped as acid formation stage. Micro-organisms taking part in this phase are termed as acid formers. As a group, these organisms are rapidly growing and are not much dependent upon surroundings. Products of first two stages serve as the raw material for the third stage where organic acids are utilized as carbon source by Methane forming micro-organisms, which are also known as Methanogens. These Methanogens are more susceptible to their surroundings. The tolerated pH range is 6.8 to 7.5 with optimum 7.0. Any departure from this range is inhibitory. Atmospheric Oxygen is extremely toxic for methanogens, as they are strict anaerobes.

12.3 Parameters Affecting Anaerobic Digestion

There are several parameters which affect the anaerobic digestion/ gas yields and they can be divided into two parts:

12.3.1 Environmental factors

There are a few environmental factors which limit the reactions if they differ significantly from their optimum levels. Factors of most interest are (a) temperature, (b) pH and (c) nutrient contents of the raw materials.

(a) Temperature

It is a factor which affects most small & medium size biogas installations in developing countries. There are three zones of temperature in which biogas is produced by anaerobic fermentation of organic matter, viz.: 1) Mesophillic 2) Thermophillic and 3) Psycrophillic zones. The optimum temperature of digester slurry in Mesophillic zone is 35°C, 55°C in Thermophillic zone and 10°C in Psycrophillic zone. In different temperature zones different sets of microbes, (bacteria) especially the methanogens remain active; whereas the other two groups of microbes either remain dormant and thus more or less inactive as far as the anaerobic digestion is concerned or get killed. However, the rate of fermentation is much faster at high temperature. Most rural household biogas plants (digesters) in developing countries operate at ambient temperatures, thus digester slurry temperature is susceptible to seasonal variation but is more dependent on the ground temperature than the atmospheric temperature. As a result, gas output in winter falls by up to 50 per cent. Below a slurry temperature of 10°C all the reactions cease to take place but revive gradually with the rise in temperature.

(b) pH

The pH range suitable for gas production is rather narrow i.e. 6.6 to 7.5. Below 6.2 it becomes toxic. pH is also controlled by natural buffering effect of NH_4^+ and HCO_3^- ions. pH falls with the production of volatile fatty acids (VFAs) but attains a more or less constant level once the reaction progress.

(c) Nutrient Concentration

Biogas producing raw materials can be divided into two parts i.e. 1) Nitrogen rich and 2) Nitrogen poor. Nitrogen concentration is considered with respect to carbon contents of the raw materials and it is often depicted in terms of C N ratio. Optimum C/N ratio is in the range of 25 to 30:1. In the case of cattle dung the problem of nutrient concentration does not exist as C/N ratio is usually around 25:1.

12.3.2 Operational Factors

Operational factors contributing to the gas production process are: (a) Retention time (RT) - also referred as detention or residence time, (b) Slurry concentration and (c) Mixing.

(a) Hydraulic Retention Time (HRT)

The number of days the feed material is required to remain in the digester to begin gas production – is the most important factor in determining the volume of the digester which in turn determines the cost of the plant; the larger the retention period, higher the construction cost. In India, the different HRTs are recommended for three different temperature zones as shown in Table 12.1.

Table 12.1 Different HRT Zones

Zone	Average ambient temperature	HRT (days)	Approximate regions
I	>20°C	30	Andhra Pradesh, Goa, Karnataka, Ketala, Maharashtra, Tamil Nadu, Pondicherry and Andaman & Nicobar Islands
II	15-20°C	40	Bihar, Gujrat,Haryana, Jammu region of J&K, Madhya Pradesh, Orrisa, Punjab, Rajasthan, Uttar Pradesh and West Bengal
III	<15°C	55	Himanchal Pradesh, North-eastern states, Sikkim, Kashmir region of J&K, and hill districts of UP

(b) Slurry concentration

This is denoted by dry matter concentration of the inputs. The optimum level for cattle dung slurry in the range of 8 to 10% and any variations from this result in lower gas output. Mixing four parts of dung with five parts of water forms a slurry with dry matter concentration of about 9 per cent, whereas 1 part of dung to 1 part to water would give a slurry concentration of 10 per cent. This also affects the loading rate etc.

(c) Mixing & Stirring

Proper mixing of manure to form an homogenous slurry before it is fed in the digester, is an essential operation for better efficiency of biogas systems; whereas proper stirring of digester slurry ensures repeated contact of microbes with substrate and results in the utilization of total contents of the digesters. An extremely important function of stirring is the prevention of formation of scum layer on the upper surface of the digester slurry which, if formed, reduces the effective digester volume and restricts the upward flow of gas to the gas storage chamber. Mixing results in premature discharge of some of the input and a perfectly unmixed system is likely to result in better reaction rate but for the problem of scum formation.

12.4 Constituents

The gas thus produced by the above process in a biogas plant does not contain pure methane and has several impurities. A typical composition of such gas obtained from the process is as given in Table 12.2.

Table 12.2 Constitutes of Biogas

Methane	60.0%
Carbondioxide	38.0%
Nitrogen	0.8%
Hydrogen	0.7%
Carbon-monoxide	0.2%
Oxygen	0.1%
Hydrogen Sulphide	0.2%

The calorific value of methane is 8400 kcal/ m^3 and that of the above mixture is about 4713 Kcal/ m^3. However, the biogas gives a useful heat of 3000 kcal/m^3.

12.5 Calorific Value

The calorific values or heat values indicate (Table12.3) that biogas can perform works similar to fossil oil in domestic cooking, lighting etc., with better efficiency depending upon the methane content in it. The biogas has also the potential for use in internal combustion engines used for pumping water etc. for which research and development works are in progress. Biogas, therefore, has a bright future as an alternate renewable source of energy for domestic and farm use.

Table. 12.3 Calorific Value of Different Fuel

Commonly used fuels	Calorific values in Kilo calories	Thermal efficiency
Biogas	4713/M^3	60%
Dung cake	2093/Kg	11%

Contd...

Table 12.3. Contd...

Firewood	4978/Kg	17.3%
Diesel (HSD)	10550/Kg	66%
Kerosene	10850/Kg	50%
Petrol	11100/Kg	-

12.6. Pre-requisites of Biogas System

(i) Land and site

While selecting a site for a biogas plant, following aspects should be considered:

a) The land should be leveled and at a higher elevation than the surroundings to avoid runoff water.

b) Soil should not be too loose and should have a bearing strength of 2 kg/cm^2

c) It should be nearer to the intended place of gas use.

d) It should also be nearer to the cattle shed/ stable for easy handling of raw materials.

e) The water table should not be very high.

f) Adequate supply of water should be there at the plant site.

g) The plant should get clear sunshine during most part of the day.

h) The plant site should be well ventilated as methane mixed with oxygen is very explosive.

i) A minimum distance of 1.5m should be kept between the plant and any wall or foundation.

j) It should be away from any tree to make it free from failure due to root interference.

k) It should be at least 15m away from any well used for drinking water purpose.

l) There should be adequate space for construction of slurry pits.

(ii) Feed for gas plants

The feed for gas plants in India mainly comprises of dung from cattle. Although, quantity of dung per cattle depends upon health, age, type and many other factors, it is generally believed that, average cattle yield is about 10 kg dung per day. On this presumption, the number of cattle required for various sizes of gas plants as has been recommended by KVIC can be taken from following table.

(iii) Temperature

Temperature plays the most important role in the bio-gas production. The total amount of gas production from a fixed weight of organic waste is best when the temperature is within the messophillic range 25-37°C and thermophillic range between 45-55°C. The gas yield is maximum in the thermophillic region and the period of digestion is also reduced. It takes about 55 days in messophillic range for digestion where as it takes about 7 days in thermophillic region.

(iv) Hydrogen ion concentration

pH of slurry in the digester should be maintained between 6.8 and 7.2 for optimum gas production and this can be accomplished by maintaining proper feeding rate. pH indicates the acidity and alkalinity of the feed mixture. Any excessiveness of acidity or alkalinity would affect gas production. There are a number of ways to correct the pH if the slurry becomes acidic or alkaline.

(v) Agitation

Mechanical agitation of the scum layer and slight stirring of slurry improves gas production but violent stirring retards it.

(vi) Solid content

The solid content in the slurry should be maintained between 7.5 to 10 per cent for optimum gas production.

(vii) Carbon to nitrogen ratio

A carbon to nitrogen ratio of 20: 1 to 30: 1 is found to be optimum for bio-gas production. Carbon to nitrogen ratio of various materials is given in Table -12.4 as a guide so that the C: N. ratio of bio-gas feed mixture is kept at desired level.

Table 12.4 Carbon to Nitrogen Ratio of Various Materials

S. No.	Material	Nitrogen Content (%)	Ratio of Carbon to Nitrogen
1.	Urine	15.18	8:1
2.	Cow dung	1.7	25:1
3.	Poultry manure	6.3	N.A.*
4.	Night soil	5.5-6.5	8:1
5.	Grass	4.0	12:1
6.	Sheep waste	3.75	N.A. *
7.	Mustard straw	1.5	20:1
8.	Potato tops	1.5	25:1
9.	Wheat straw	0.3	128:1

12.7. Main Features of the Biogas Plant

On the basis of the gas holder the present biogas plants are classified mainly into two groups -fixed dome type or floating drum type. Both the type of plants have the following functional components:

(i) *Digester* : This is the fermentation tank and is built partially or fully underground. It is generally cylindrical in shape and made up of bricks and cement mortars. It holds the slurry within it for the period of digestion for which it is designed.

(ii) *Gas holder:* This component is meant for holding the gas after it leaves the digester. It may be a floating drum or a fixed dome on the basis of which the plants are broadly classified. The gas connection is taken from the top of this holder to the gas burners or for any other purposes by suitable pipelines. The floating gas holder is made up of mild steel sheets and angle iron and is required to exert pressure of 10 cms of water in the gas dome masonry and exert a pressure upto 1m of water column on the gas.

(iii) *Slurry mixing tank:* This is a tank in which the dung is mixed with water and fed to the digester through an inlet pipe.

(iv) *Outlet tank and slurry pit:* An outlet tank is usually provided in a fixed dome type of plant from where slurry in directly taken to the field or to a slurry pit. In case of a floating drum plant, the slurry is taken to a pit where it can be dried or taken to the field for direct applications.

12.8 Classification of Rural Household Digester

There are three basic methods by which rural household biogas digesters in developing countries are operated in practice, namely: (i) Batch, (ii) Semi-continuous and (iii) Semi-batch digesters.

(a) Batch digester

In this process the whole digester is filled with raw materials for gas production along with some starting (seed) material. This is allowed to ferment and produce gas over a certain length of time and when gas yields become very low the digester is emptied of all the sludge which can be supplied as manure. In this system gas production begins at a low level and goes on increasing only to drop down again after reaching the peak. Because of variable gas production level, high cost and periodic emptying and filling of digesters, this process has not become popular. Examples of these digesters are small size garbage plant and crop-residues plant.

(b) Semi-continuous digester

The rural household digesters are fed once a day and the fresh input displaces the same volume of spent materials from the digester. Every day a certain quantity of fresh inputs is fed into the digesters which is expected to remain in the digester for a prescribed retention time and produces gas over this length of time before being discharged.

(c) Semi-batch digester

A combination of batch-fed and semi-continuous fed digestion is known SHF digestion. Such a digestion system is used where the waste like garbage etc., which are available on daily or weekly basis but cannot be reduced to make slurry. In the semi-batch system, the animal manure can be added on a daily basis after the initial loading is done with garbage, agricultural waste, leaves, crop residence and water hyacinth etc.

12.9 Size Selection of Rural Household Biogas Plants

Size of the rural household biogas plant to be installed should be selected on the basis of gas requirement and the livestock manure availability with the beneficiaries. Since cattle dung is the main substance for the biogas plant in rural India, the table given below shows the relationship between plant capacity, daily cattle dung requirement and gas use (Table-12.5).

Table 12.5. Relationship between plant capacity and daily dang required.

Size No	Plant capacity m^3	Daily dung Required (kg)	Approximate No. of cattle	No. of family members
1	1	25	2 - 3	3 - 4
2	2	50	4 - 6	5 - 8
3	3	75	7 - 9	9 - 12
4	4	100	10 - 12	13 - 17
5	6	150	12 - 20	18 - 25

12.10 Popular Designs of Biogas Plant Models

There are three popular Indian designs of biogas plants namely: KVIC, Janata and Deenbandhu biogas plants. For construction of KVIC & Janata model plants - Indian Standard IS:9478-1986 released by Bureau of Indian Standards should be followed. Brief description of the three models is given below.

(a) KVIC plant

It was in or around the year 1945 that Scientists at Indian Agricultural Research Institute (IARI), New Delhi worked on semi-continuous flow digesters and in the year 1961 Khedi and Village Industries Commission (KVIC) patented a design which is being popularized by various agencies in many countries. This design consists of a deep well shaped underground digester connected with inlet and outlet pipes at its bottom, which are separated by a partition wall dividing the ¾th of the total height into two parts. A mild steel gas storage drum is inverted over the slurry which goes up and down around a guide pipe with the accumulation and withdrawal of gas. Now FRP and ferro-cement gas holders are also being used in the KVIC plant (Fig. 12.1).

(b) Janata plant

The Janata model is a fixed roof biogas plant which was developed by PRAD in 1978. This is also a semi-continuous flow plant. The main feature of the Janata design is that the digester and gas holder are part of a composite unit made of bricks and cement masonry. It has a cylindrical digester with dome shaped roof and large inlet and outlet tanks on two sides. It requires shuttering for making the dome shaped roof and a skilled & trained master mason is a must for the construction of a Janata Biogas plant. This plant costs about 20-30% less than the KVIC floating drum type plant (Fig. 12.2).

(c) Deenbandhu plant

Unlike Janata biogas plants, for constructing plants of this design no shuttering is required for making the dome shaped roof. This also results in less labour and time required for completing the construction (Fig. 12.3).

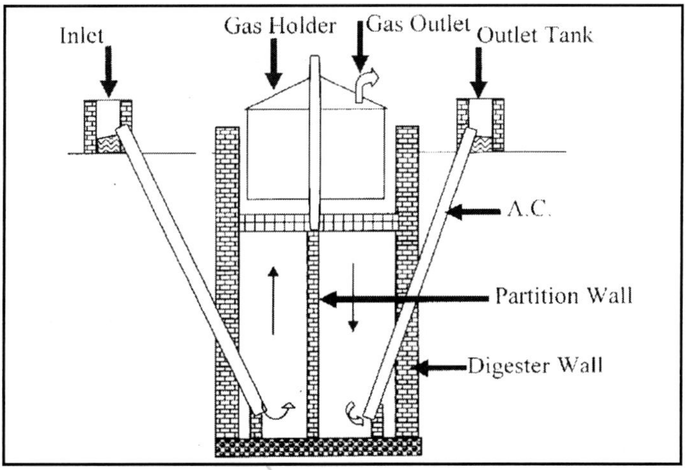

Fig 12.1. KVIC Biogas Plant.

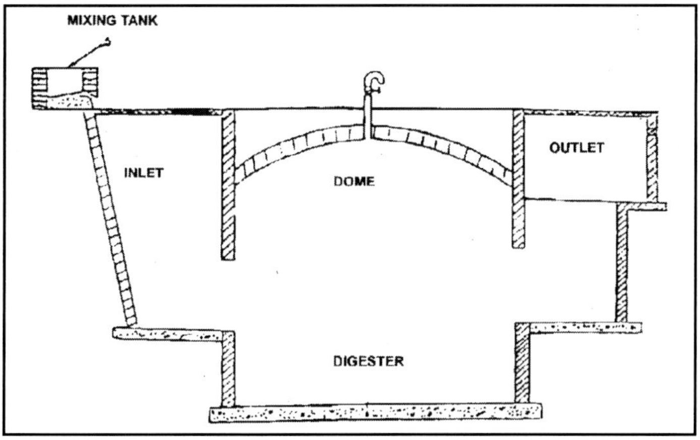

Fig 12.2. Janta Biogas Plant

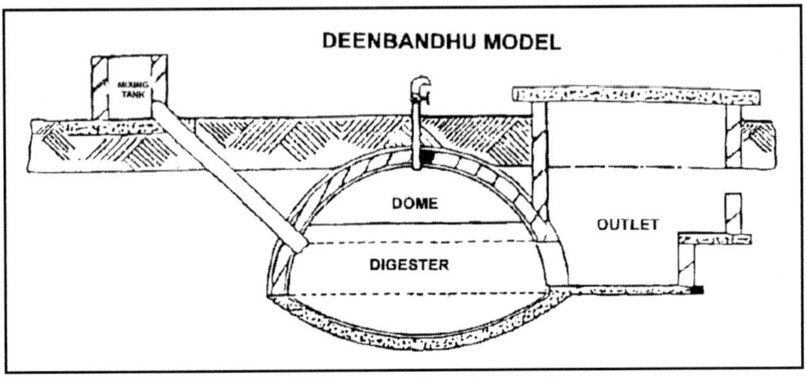

Fig 12.3. Deenbandhu Biogas Plant

12.11 Comparisons among Family Size Biogas Plants

#	KVIC	Janata	Deenbandhu
1.	The digester of this plant is a deep well shaped masonry structure. In plants above $3m^3$ there is a partition in the middle of the digester.	Digester of this plant is a shallow well shaped masonry structure. No partition wall is provided.	Digester is made of segments of two spheres one each for the top and bottom.
2.	Gas holder is generally made of mild steel. It is inverted into the digester and goes up and down with formation and: utilization of gas,	Gas holder is integral part of the masonry structure of the plant, Slurry from the gas storage portion is displaced out of the digester with the formation of gas and comes back when it is used.	The structure described above includes digester and the gas storage chamber, Gas is stored in the same way as in the case of Janata plants.
3.	The gas is available at a constant pressure of about 10cm of water.	Gas pressure varies from 0 to 90 cm of water.	Gas pressure varies from 0 to 75 cm of water.

Contd...

4.	Inlet and outlet connections are provided through A,C pipes	Inlet and outlet tanks are. large masonry structures designed to store the slurry displaced out of the digester.	Inlet connection is through AC pipe. 0utlet tank is large masonry designed to store displaced slurry.
5.	The volume of the gas holder governs gas storage capacity of the plant.	.it is the combined volumes of inlet and outlet chambers, (portions of inlet and outlet tanks above the second step.)	It is the volume of outlet displacement chamber
6.	The floating mild steel gas holder need regular maintenance to prevent corrosion. It has short life.	There is no moving pert and hence no recurring expenditure. It also has e long working life.	There is no moving part and hence no recurring expenditure, It also has a long working life.
7.	Installation cost is very high, A 2 m^3 plant costs over Rs. 8000.00.	It is cheaper than the KVIC type plants, A 2 m^3 plant costs about Rs. 5500.00.	It is much cheaper then KVIC and Janate type plants. A 2 m^3 'plant of this design costs Rs 4200.00.
8.	Digester can be constructed locally the gas holder needs sophisticated workshop facilities.	A trained mason using locally available materials can build entire plant.	Entire plant can be built by a trained. mason using locally available materials.

12.12 Major Benefits of Installing Biogas Plants

It is estimated by the Govt. of India, Ministry of Energy, that alternative sources of energy like bio-gas plants, wind mills etc. may reduce the dependence on conventional sources of energy by about 20 per cent by the turn of the century, provided promotional efforts are continued. Presently, the cooking media in rural areas consist of burning dung cake, fire-wood and to some extent kerosene where it is available easily. The installation of biogas plants would directly replace the use of above three and in saving them, following gains would be made:

(i) Nearly 30 per cent of available dung which is burnt and wasted would be recovered as bio-gas plants conserve the dung while producing bio-gas. Again, the dung after digestion in gas plant preserves more of NPK in the dung solids and cellulose which otherwise gets lost if heaped in the open.

The benefits derived from bio-gas plants in terms of manure and useful energy are illustrated at Annexure 1& II. The average NPK content of Farm Yard Manure (FYM) is about 0.5, 0.2 and 0.5 per cent respectively and it may be observed that biogas slurry is rich in NPK by more than four times than ordinary dung when converted into FYM. When the country is faced with shortage of fertilizers and has to spend enormous amounts for its import, the application of bio-gas slurry can replace the chemical fertilizers to a large extent. Bio-gas slurry or FYM not only adds NPK but it proves the soil porosity and texture. These are established benefits.

(ii) Second major benefit is that rural people would gradually stop felling trees. Tree felling bas been identified as one of the major causes of soil erosion and worsening flood situation. Government has started massive afforestation programme to tackle the erosion and flood situation. Continued deforestation has been causing ecological imbalances in the environment in which we live. Biogas plants would be helpful in correcting this situation.

(iii) In rural areas kerosene is used for lighting lantern and cooking in a limited way wherever kerosene supply has been made possible. Whatever quantity is used can be replaced by biogas as it can be used for lighting and cooking. This would reduce the dependence on fossil oil directly and in saving foreign exchange.

(iv) Lastly, the most important social benefit would be that the dung being digested in the digester, there would be no open heap of dung to attract flies, insects and infections. The slurry from digesters can be transported to the farm for application in the soil, thus keeping the environment clean for inhabitation. Also, gas cooking would remove all the health hazards of dung cake or fire wood cooking and would keep the woman folk free from respiratory and eye diseases which are prevalent in the villages.

Equivalent Quantity of Fuel for 1 m^3 of Bio-Gas

Name of the fuel	Equivalent quantities to 1 m^3 of Bio-gas
Kerosene	0.620 kg
Firewood	3.474 kg
Cowdung cakes	12.296 kg
Charcoal	1.458 kg
Soft coke	1.605 kg
Butane	0.433 kg
Furnance Oil	0.4171
Coal gas	1.177 m^3
Electricity	4.698 kWh

12.13 Biogas Application and Appliances

Cooking and Lighting: The main use of tbe bio-gas in rural areas is cooking. Firewood, crop residues or dung cakes are not available on regular basis and so are other conventional sources of energy like kerosene, electricity or coal etc. The other 'use is gas lamp which glows like any bright lamp. Generally, 15 running meter length (RML) of pipes/tubing are allowed in biogas plant schemes along with burners and lamps for schemes financed by banks.

(i) Dual fuel engines

This is a recent appliance where certain modification of air intake system helps carburation of biogas to run the diesel engines. It is well known that diesel engine has wide application in rural areas from irrigation to any stationary operations and these engines can be converted to dual fuel ones. This dual fuel engine is in a position to make use of about 70 per cent biogas and 30 per cent diesel. The economy of running dual fuel engine by bio-gas is undisputable but it has certain operational difficulties. The main problem is that while biogas plants are located near the house, the running of the diesel pumpsets are required in the fields. As such it is impractical and technically unsound to provide long gas pipes to connect the gas plant with the diesel engines in the field. Another aspect that prohibits the use of bio-gas in diesel engine is that it requires larger gas plants at least 8-10 cum, so as to enable a 1ow 3 HP diesel engine to run for 4-5 hours a day. And often the users prefer the sizes of 2-4 cum gas plants as number of cattles are not many. However, considering large bio-gas plants hold the key to economical operation of dual fuel engines, in future one may see its wide spread application when many such gas plants come into existence.

(ii) Research and development in application of biogas

Unless ways and means are found for diverse application of biogas, it may not receive the appreciation in deserves. Community bio gas plants have been mooted to provide cooking gas, street light and drinking water through pumping. Government of India had proposed during 1983-84 to install 500 community gas plants on pilot basis and study their feasibility for running on cooperative basis. The results may be helpful in spreading similar plants to other areas. Besides, many institutions/individuals in the country have attempted to provide alternative gas plant designs and some 30 designs have been tried so far. Yet the KVIC design and Janata Model are the two which are not only but are recognized by the financing institutions as standard designs and this status would be maintained unless better designs come as replacements. In certain parts of the country use of night soil and connecting latrine to bio-gas plants have already found its social acceptability. However, in most part of the country it is still not socially accepted and when it finds its social acceptability, it would ensure a regular supply of feed besides improving sanitation in rural areas.

13

Harnessing the Power from Wind

13.1 Introduction

Winds are the motion of air around the earth. This movement in air is caused by the uneven heating of the planet's surface by the sun. The idea of using wind as a form of power is not new. The traditional applications of wind were primarily as sources of kinetic energy for rural, agricultural and a limited number of industrial uses such as pumping water and grinding grain.

Presently wind energy can be used for two major applications, such as wind mills for pumping water for drinking as well as for irrigation purposes and second application is as aero-generator for electricity generation for domestic and industrial uses. In addition to this, presently wind energy battery charger are also available, which can store energy for lights, radio communication, hospital equipment and to power various emergency related equipment.

The wind energy has an enormous resource, but the problems of utilizing the winds are many and varied. Wind energy is very diffuse in nature and local topographical features significantly alter the prevailing winds thus leading to the extremely site specific nature of wind energy.

Scientirsts have estimated that as much as 10 per cent of the world's electricity could be provided by wind generators by the middle of 21 century. India has estimated wind power potential of 45,000 MW. India now, ranks fourth in the world in wind power generation.

The world's largest wind farms are in California, where wind turbines can generate power upto about 1120 MW. The first wind mill used as source of electric power was built in Denmark in 1890. Presently in India about 2382 MW of electricity is being produced through wind energy.

13.2 Wind Energy and Its Characteristics

The energy in the wind is the kinetic energy due to stream of air particles of density (δ), moving with a velocity (v) through an area (A) swept by the rotor of diameter (D) (Fig. 13.1).

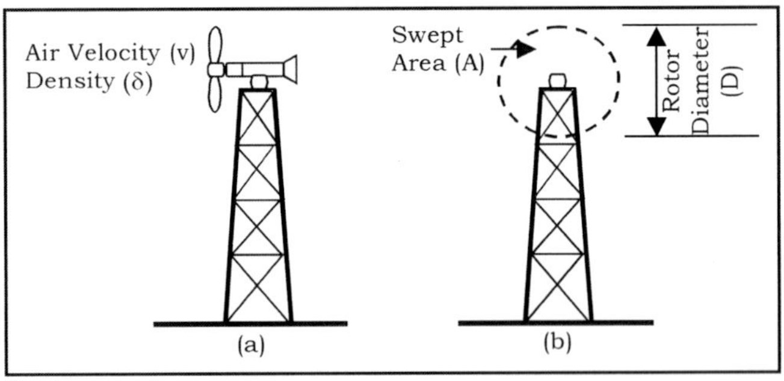

Fig. 13.1. Energy in the Wind Due to
(a) Air Velocity and Air Density (b) Swept Area

The mass flux M of the air is given by

$$M = \rho AV$$

and the power in the wind.

$$P_m = 0.5 \text{ M } V^2$$

or

$$P_m = 0.5 \rho A V^3$$

where

V = Velocity in m/sec,

A = Exposed area in m^2,

ρ = Density of the air in kg/m^3. The density is the function of pressure, temperature and relative humidity, and

P_m = Power of wind in Watt.

The theoretical maximum power that may be captured by a wind machine was shown by Betz in 1920 to 16/27 or 59.3 per cent of the incoming energy in the wind.

The major factor influencing energy in the wind is its velocity. As shown above, the power is a factor of the cube of the velocity, so doubling the wind speed will increase the available power eight times. However, wind energy is very diffuse in nature and also the velocities are extremely variable.

The practical output in terms of electrical energy or amount of water lifted can be calculated by using following equations

Power = $0.5\ A\ \rho\ v^3.C_p$ watts

where

V = Average wind speeds m/s

A = Swept area of rotor, m^2

C_p = Ratio of available power extracted to the theoretical power available

In case of wind will C_p = 0.4 and generating efficiency = 0.8

Thus effective power = $0.19\ A\ v^3$ watts

In case of Aero generator, Effective power = $0.2\ Av^3$ watts

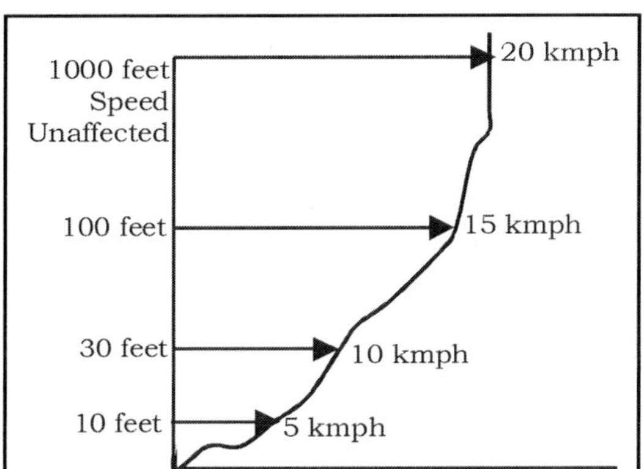

Fig. 13.2. Effect of Surface Roughness on Wind Velocity

The energy output is found multiplying with the duration of the period concerned. This is calculated for a year showing the annual output as a function of the average wind speed (v).

The efficiency of wind machine depends on wind speed, with type of turbine and with the nature of load. However, the wind speed is playing an important role. As the wind increases from a low value, the turbine is able to overcome all mechanical and electrical losses and start delivering electrical power to the load at cut in speed V_c. The rated power out put of the generator is reached at rated wind speed V_R. Above that speed, some wind power is spoiled to maintain constant power output. At the furling speed V_F, the machine is shut down to protect is from high winds. V_R can be increased by using a larger generator for a given wind turbine. This increases the static friction so that V_c will also increases. The cut in speed will typically be about one half the rated wind speed, which corresponds to cut in power of one eight the rated power. Below V_c most of power in the wind is used to overcome mechanical and electrical losses.

Wind is highly variable source and there are several methods of characterizing this variability. Most common is the power duration curve to select V_c and V_R for a given wind site. Seasonal and diurnal variation has significant effect on wind. Load duration data are required to judge the appropriate height. Diurnal variation is less with increase in heights. Due to seasonal variation average power may typically vary from 65 to 125 per cent. This factor must be suitably considered in the design parameters.

Wind speed increases with the height because of friction at earth surface (Fig. 13.2). The rate of increase is given by

$$V/V_O = (H/H_O)^{1/7}$$

Where V is the predicted wind speed at height H and V_O is the wind speed at height H_O. This translates into substantial increase in power at greater heights.

13.3 Design of Wind Energy System

The primary step of the design of wind energy system includes data collection and its analysis.

13.3.1 Data collection

As air moves across the surface of the earth, its speed and direction are changed by local topography as well as by local heating large bodies of water, or other small scale influences can disturb the flow of air. The combined effects can lead to highly variable winds and results in the site specificity of potential wind energy sites.

Topographic surface roughness increases the friction on the boundary layers of air at the surface and shearing of the boundary layers may result. Fig. 13.2 indicates this effect and the need to site windmills and the measuring equipment above the affected regions.

In areas of hilly or mountainous terrains, wind velocities and directions will be seriously altered by local topography. Such features affect airflows on much larger scale than surface roughness, but their combined effects should always be taken into account. Non flat terrain can be divided into two categories:

(i) Elevated terrain

(ii) Depressions

(i) Elevated terrain

The main topographical features associated with elevated terrain are ridges, hills and cliffs. Their effects are generally two fold, namely:

1. The elevated terrain can act as huge towers, thereby exposing a windmill to winds of high velocities.
2. Airflow can be accelerated when moving around elevated terrain, thus enhancing the wind velocities.

However, elevated terrain can also have adverse effects, as slowing of the winds, turbulence and wind shear can occur, but proper site analysis can eliminate these hazards.

(ii) Depressions

Terrain features such as valleys, basins, gorges and passes are associated with depressions. Generally, depressions will lead to areas which are sheltered from the prevailing winds and are most likely to be affected by the diurnal changes in wind patterns. While some valleys may have a funnelling effect on the winds and increase their velocities, depressions do not usually offer good sites for wind energy.

13.3.2 Data analysis

The most commonly applied technique for assessing wind energy is to first tabulate the available data, and then calculate monthly, daily and, in some cases, hourly average wind speeds. These mean wind speeds are plotted on a graph of "Mean Wind Speed vs. Time", or densities (i.e. power per unit area) and the results then analysed. Integrating these figures over the particular time period gives the power densities available over that time period.

However, when considering a wind energy conversion system (WECS) a cut in speed, rated speed and cut out speed must be considered. Analysing the data along with the specifications of different WECS is much more useful, and the actual energy output from the WECS may be calculated for the time period over which the data was taken.

There are at least two major constraints on wind turbine design one is to maximize the average power output. The other is to meet the necessary load factor (which is the ratio of average electrical power to the rated electrical power) requirement of the load. Load factor is not of major concern if the wind electric generator is acting as a fuel saver on the electric network. But if the generator in pumping irrigation water in a synchronous mode for example, load factor in very important.

Design for wind turbines have been driven by three basic design concepts for handling wind load i.e.

(1) Withstanding the load
(2) Shedding or avoiding of loads
(3) Managing load mechanically and/or electrically.

Important characteristics of first design concepts are optimization for reliability, high solidity but non optimization blade pitch, low tip speed ratio (TSR), three or more blades. Turbines based on second concept have design criteria like optimization for performance, low solidity, optimum blade pitch, high TSR/optimum blade pitch etc. where as in third concept design considerations like optimization for control, two or three blades, moderate TSR, mechanical and electrical innovations (flapping or hinged blades, variable speed/low speed generators).

13.4 Wind Measurement Equipment

The type of equipment to be used for estimating the power in the winds depends largely on the type of data required. If the WECS is to be used for estimating annual power output, mean wind speeds and wind speed distributions for one year are required. If the WECS is to be used for providing power to match peak power periods more detailed records of the diurnal changes are required. For small power applications (e.g. pumping of water) mean annual or monthly wind speeds are sufficient.

Wind measurement apparatus consist of wind sensors and displays or recorders. The wind sensors are cup or propeller anemometers which sense wind speed and wind vanes for sensing wind direction.

The wind sensors usually produce an electrical signal, but in some cases it is mechanical, which is monitored by a recorder. These recorders may be counters, strip chart recorders, or magnetic tape recorders and provide a record of the winds which can be analysed. The simplest recorder is the wind run anemometer which has a mechanical counter measuring the amount of wind passing the sensor. For obtaining average wind speeds this system is adequate, but if a more detailed record is required, either the strip chart or magnetic tape recorders must be used. However, strip chart and magnetic recorders are more expensive and delicate instruments.

Strip chart recorders require manual tabulation of the recorded data, and this can be a tedious and inaccurate process. Magnetic tape recorders, on the other hand, are easily processed with computers, but such facilities must be available.

In selection of the wind measurement sites, attention should be given to the various topographical factors that affect wind speeds. The anemometers should be placed at height where they will be least affected by the local environment, or at the height where they will be least affected by the WECS, if a commercial system is being considered.

Where ever possible, wind data available from a local weather station should be consulted, and some kind of correlation made between the existing data. If a good correlation exists, then some indication of how typical the data is for that period will be obtained and greater assurance in the accuracy of the data will then exist.

13.5 Wind Energy Conversion Systems

The use of crude windmills to pump water and to power simple machines dates back a long time. A picture of windmill in a Chinese vase of about the third millennium BC is the earliest, but the Persians were using windmills for grinding grain by seventh century AD Until the early twentieth century, windmills were in common use for providing electricity, on a small scale, and for pumping water in rural areas. In addition, the wind was the motive power for the European colonization of most of the world in the sixteenth and seventeenth centuries. In more recent times, the windmill is being assessed as a potential source of power generation both on large and small scale. In general, WECS are divided into two types, horizontal axis machines and vertical axis machines, some typical examples of both types are shown in Fig. 13.3. The horizontal axis machines are those in which the axis of rotation is parallel to the direction of the wind. These WECS suffer the disadvantage of having to be oriented into the wind when the wind direction changes and in small machines this is usually accomplished by a simple tail vane held by spring tension. In large horizontal axis WECS, sophisticated electronic sensors and motors may be required to achieve the desired orientation.

Vertical axis machines have their axis of rotation at right angle to both the Earth's surface and the direction of the wind and they have the advantages that do not have to be repositioned in to the direction of the wind.

The simplest vertical axis WECS is the Savonius type rotor, which is high torque, low speed machine that can be constructed very cheaply from old drums (see Fig. 13.3) This machine is vary suitable for pumping water and generating small quantities of electricity (kW). The efficiency,

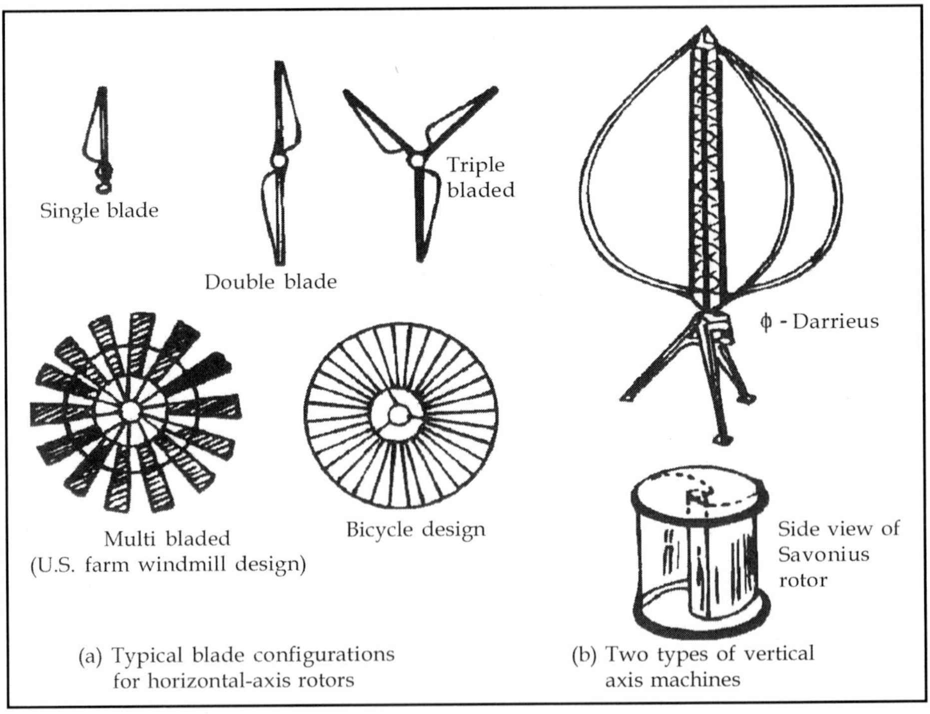

Fig. 13.3. Wind Energy Conversion Systems

however, vary at about 10-15 per cent depending on the wind speed, so far larger requirements, the Savonius is not recommended. Larger vertical axis rotors such as the Darrieus turbines are being developed extensively for large power generating systems, but their applicability may be hindered by the need for high average wind speeds (24 mph) and an inherent efficiency which is lower than from horizontal axis propeller machines.

The advantages and disadvantages of horizontal axis and vertical axis machines are given in Table 13.1. The earliest wind mills were of the fixed blade variety but to improve efficiency, particularly of large WECS, pitch modulations mechanisms are now being utilized. These mechanisms change the angle of attack of the blades according to changes in wind speed. This results in additional expense to the system, but the increased efficiency at low wind speeds usually offsets this cost disadvantage.

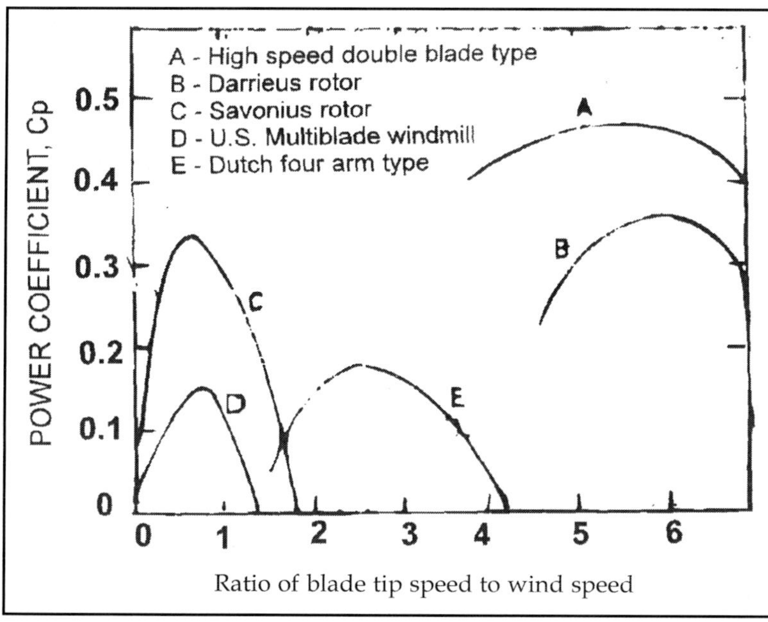

Fig. 13.4 Typical Performance Curves for Various Wind Turbines

The Fig. 13.4 shows some performance curves for different WECS. The power coefficient, C_p is a measure of the efficiency of the wind turbine and it is seen that A, the high speed double blade type horizontal axis rotor is the most efficient. For large power generating systems, this is the design that is now achieving the most interest. The simple Savonius rotors and US multiblade windmill are low efficiency machines but for meeting small power requirements they are very well uited as they are the most inexpensive machines and may frequently be fabricated with a minimum of sophisticated equipment.

Table 13.1: Comparison of Horizontal & Vertical Axis Machine

S. No.	Type of Machine	Advantage	Disadvantage
1.	Horizontal axis Machine	Rotor weight in less	Pitch changing mechanism for speed control requires maintenance
		High speed are possible	Tower required for mounting
		Solidity ratio are less	Up wind rotors should have enough clearance from the tower to avoid striking
		Large experience available	Down wind rotor required for large sized
		Available in sub kw to MW range	Directional control is necessary
2.	Vertical axis machine	Less land for structure	Solidity ratio is high
		Directional control not required	Rotor is heavy
		Surrounding of blades can be used against gusts	Suitable for small power generation
		High rpm is possible	High starting torque required

13.6 Wind Energy Devices

The wind energy is successfully used for water pumping, grinding of grains and for electricity generation. The details of devices available in the field in India are as follows:

13.6.1 Wind pump

The size of wind pumps currently available range from 1 to 8m rotor diameter. The pumping height and average wind speed influenced the average power output which can range from a few watts to about 1 KW. For higher power demand wind electric pumping systems (WEPS) can be applied, incorporating a wind generator (available in larger diameter) driving an electronic motor pump combination. At higher power output levels (10 KW or more), the pumping system could be integrated with a small electric grid supplying electricity for other purposes besides water pumping. For average rural Indian applications, the daily output ranges from 1.5 to 500 m^3 at a head of 20 m or roughly from 0.1 to 3 kWh/day. The reasonable costs of pumping water run upto Rs. 20 per m^3. The pumps are available which can be operational at wind speeds of 5-6 km per hour or more.

13.6.2 Aero generator

Wind energy can be successfully tapped for generation of electricity. Currently wind generated power accounts for about 4500 MW of installed capacity world wide. In India, an aggregate capacity of more than 2382 MW has already been installed and an addition of 500 MW per year is envisaged.

Wind farms, which constitute a cluster of grid connected wind electric generators of 250-500 KW aero generators, have proved to be a feasible method of power generation on a large scale. It is well suited in locations where the annual average wind speed is at least 18 kmph or the energy content is above 2000 kwh/m^2/year. As wind farms need large open spaces (8-10 hectares/MW) the availability of land area in windy region is extremely important. But within a wind farm, only 5 to 10 per cent of the land area is actually required for installation of the wind electric generators and roads etc. the remaining land can be put to productive use such as crop growing, cattle grazing etc.

The average capital cost of a wind farm project works out to Rs. 3.5 to 4.0 crore per MW which is fairy comparable with cost of setting up a conventional power plants. Based on this investment and normal methods of cost calculation the average cost of electricity generation in wind power projects is in the range of Rs. 2.25 to 2.27 per unit, depending upon the site.

Government of India has provided provision of subsidy for setting up of wind farm commercially and also for specific categories of users. Commercial organisations and professional using wind electricity for their on going plant can claim 100 per cent depreciation in the first year of the installation. Electricity for their on going plant can also be availed from Indian Renewable Energy Development Agency for setting and to promoting manufacture and utilisation of wind energy gadgets. The power generated from the wind farm can be sold to SEBS or transmitted through SEB lines to the place of use by the owner of the wind energy based power plants or sold to third party.

The import of wind energy based gadgets are also easy such as import of aero generator, control systems and materials are covered under the open general license. Further, imports at concessional rates of custom duty are permitted to the manufactures of aero generators in the country.

13.6.3 Wind battery charger

Wind energy generators can also be used to charge batteries, which can store energy for lighting, radio communication, hospital equipment and to power various emergency related equipment. These wind energy generators are known as wind battery chargers. Battery chargers are wind machines generally with a rotor diameter of 1 to 2.5 m and an output of 50-500 W at a 10 m/s wind speed. The components include a wind powered generator, a convertor and a container for the batteries. Presently, battery charger of rating 1.5 to 4 KW are available.

13.7 Wind Energy Generation Forecasting

The intermittence of wind energy presents a special challenge for utility system operations. While conventional power plants produce a near constant output - barring rare emergency outages - the output of a wind plant fluctuates. In some parts of the India, such fluctuations can amount to several hundred KW in a matter of an hour or two. To the extent the fluctuations cannot be predicted, they create costs for the electricity system and consumers as well as potential risks to the reliability of electricity supply.

One of the principal mechanisms a grid operator, such as an Independent System Operator (ISO), uses to limit unexpected changes in plant output are to charge suppliers a fee for "uninstructed deviations" between their forward schedules - i.e., their predicted output - and actual generation. This policy encourages suppliers to maintain a high level of reliability while also compensating the system for the costs of having either excess or insufficient generation. Typically, the fee is assessed on the deviation in each hour based on the market-clearing price of the Real Time Market. However, considering the volatility of wind plant output, some grid operators recognize that wind energy suppliers could be severely penalized if required to pay for deviations on an hourly basis, and that this would undercut the ability of a state or region to meet its future electricity needs in an environmentally sound manner.

The performance requirements for a forecasting service are dictated by the needs of both the grid operator and the wind generators. From the perspective of wind generators, the priority is to minimize the deviation between forecasted and actual plant output. For an ISO, there are two additional and more demanding priorities. The first is to anticipate changes in wind production as accurately as possible in the very short term (up to a few hours ahead). This is to enable an ISO to manage its grid operations and reserve capacity purchase decisions in an optimal fashion. For this purpose it is natural to consider persistence-type methods. However such methods are inherently limited in that they cannot predict changes in plant output that depart radically from recent trends, as might occur, for example, because of a passing weather front. Thus, in order to achieve the highest possible accuracy the methods should incorporate other data that may signal future trends, such as a conventional weather forecast or data from meteorological stations that lie "upstream" of the wind parks.

The second priority is to forecast the wind generation for the next day. The presumed goal is to enable an ISO to schedule reserve capacity as efficiently as possible. If so, then it may be less important to accurately forecast the *timing* of changes in wind generation than it is to forecast the *minimum* wind plant output during the peak load hours.

In general, a high degree of reliability and accuracy is required by ISOs and utility systems for wind plant forecasts. This is consistent with the usual high standard of reliability applied to all utility system operations. It is particularly important for the next-hour forecasts, because their accuracy declines relatively quickly the older the forecast becomes. The accuracy of next-day forecasts, in contrast, is not as sensitive to the age of the forecast.

13.8 The Forecasting Problem

The wind energy generation forecasting problem is closely linked to the problem of forecasting the variation of specific atmospheric variables (i.e. wind speed direction and air density) over short time intervals and small spatial scales for a small volume of the atmosphere (the wind plant) for a variety of time horizons (i.e. look-ahead periods). In general, this is an enormously difficult problem because of the wide variety of spatial and temporal scales of atmospheric motion that play a role in determining the variation of the key parameters within the targeted forecast space-time volume. In order to understand the different issues involved in wind energy forecasting it is useful to divide the problem into three difference time scales:

(1) Very short-term (0-6 hrs)

(2) Short-term (6-72 hours) and

(3) Medium range (3-10 days)

The skill in very short-term forecasting is related to the prediction of small-scale atmospheric features (< 200 km in size) in the vicinity of the wind plant. The major difficulty is that very little data is typically gathered on the scale of these features. This means that it is usually difficult to define the spatial structure and extent of these features. Therefore, the only viable choice is often to infer information about these features from the time series of meteorological and generation data from the wind plant. As a result, real-time data from the wind plant is usually crucial to the performance of very short-term forecasts. In fact, the 0-6 hr time scale has been defined as the period when persistence forecasts will typically outperform wind energy forecasts derived solely from predictions of the regional atmospheric circulation (i.e. the information from the wind plant or the plant's immediate environment is more important than the regional information). Thus, the forecast standard for the very short-term forecast is a persistence forecast.

The ability to forecast the wind energy generation over short-term time scales (6-72 hours) is tied to the skill of forecasting regional scale atmospheric features. These features are often referred to as synoptic scale weather systems and are the ones that are typically depicted in newspaper and TV weather presentations. It is necessary to gather data over a large volume of the atmosphere in order to define the structure of these systems. The importance of measurements at the wind plant drastically decreases at the start of this period. The real-time plant data is able to make some contribution to forecast quality at the start of the period but it has little predictive value after about 12 to 18 hours. This is fundamentally because the information that determines the variations in meteorological parameters for time periods greater than 12 hours comes from locations that are many hundreds if not thousands of kilometers away. As a result, the forecast standard shifts from persistence to climatology (i.e. the average conditions for that location and season) during this period. A climatology forecast will typically outperform a persistence forecast for most locations after about 12 to 18 hours.

The skill of medium range (3-10 days) forecasts is typically tied to the ability to forecast continental, hemispheric and global scale atmospheric circulation systems. Of course, the regional and local features are superimposed upon these large scale features so in order to produce accurate medium range hourly predictions of wind energy production a prediction system must forecast all scales of atmospheric motion from global or hemispheric to local.

At medium range time scale it is difficult to accurately predict the evolution of specific local-area or regional features that will affect the forecast target area. Therefore, most of the forecast skill is linked to prediction of general patterns that favor above average or below average winds for a substantial period of time (a day or more). The forecast standard for this time scale is a climatological forecast.

It should be noted that the distribution of atmospheric energy across the space and time scales varies substantially by region, season and atmospheric regime. This has important implications for predictability and forecast performance. If there is very little variability over a specific time scale then the absolute forecast performance is likely to be good but there is likely to be little skill over a simple persistence or climatology forecast on that time scale. Conversely, a situation with large variability over a given time scale will often result in lower absolute performance but a higher level of performance relative to simple persistence or climatology forecasts. The point is that one must be very careful in how one evaluates forecast performance and how performance is compared from location to location and season to season.

13.9 Forecasting Tools

There are two fundamental types of tools used in the forecasting process:

(1) Data gathering and

(2) Data processing

The data gathering tools include the wide range of measuring devices that provide data to the forecast process. These include measurements made at the wind plant itself as well as those in the local-area, regional and even global environment of the plant.

The data processing tools serve to transform the measurement data into a forecast for the desired period of time. The data processing tools include physical and statistical atmospheric models as well as models of the relationship between meteorological conditions within the wind plant volume and the plant output (usually referred to as a plant output model). Except for some government agencies forecast providers typically control the data processing tools but not the data gathering tools and therefore they are typically in a position of making the best use out of whatever data is made available to them. Due to the wide range of spatial and temporal scales that determine the variations in the wind energy power generation, it is necessary to use a diverse mix of data sources and types to achieve the best possible forecast performance. For wind energy forecasting, the most fundamental type of data is the time series of meteorological parameters and power generation from the wind plant itself. The generation data can be for the entire plant or for groups of turbines within the plant. The meteorological data typically consists of wind speed and direction and sometimes temperature, pressure and humidity data from sensors on one or more "met towers" within the plant's boundaries. This data is typically gathered at the hub height of the turbines deployed in the plant. The additional detail provided by generation data by turbine group and multiple met towers can be very beneficial in developing a more precise and accurate relationship between the meteorological conditions and the plant output. The availability of this time series data alone is sufficient to make a somewhat skillful short-term forecast and at least a climatology-level forecast for the intermediate and long term However, in order to achieve a higher level of forecast skill it is necessary to utilize data from beyond the plant's boundaries. Meteorological data from in situ sensors deployed and operated by government agencies have been a traditional source of data for wind energy forecasting. These include sensors on surface-based met towers deployed mostly at airports and sensors carried aloft by weather balloons to provide information about the vertical profile of temperature, humidity, winds and pressure. The main problem with this data is that the spacing between measurements is too large (because of economic constraints) to adequately represent the small or even sometimes medium scale atmospheric features that are responsible for short-term variations in wind energy output. However, these in situ sensor networks do a better job of mapping most of the features that are responsible for most of the variability over 1 to 2 day ahead time scales. Unfortunately, there are large areas (such as the oceans) where very little in situ data is gathered because of the cost of maintaining such systems in those environments.

The expectation is that remote sensing technology will eventually overcome these limitations of data resolution and coverage. Many types of atmospheric remote sensors have been developed and some have been deployed for operational use. These include Doppler radars, wind profilers (a type of fixed position Doppler radar), lidars, sodars and satellite-based radiometers. While all of these technologies have made contributions to the atmospheric forecasting process, each has had significant limitations that have reduced their level of impact on atmospheric forecast performance. However, remote sensing technology continues to rapidly move forward and there is still an expectation that the next generation of remote sensors to be deployed during the next few years will have a greater impact on forecast performance.

Data processing tools are the other major component of the forecast process. These are the mathematical (often called numerical) models that ingest data and generate predictions. There are four fundamental types of numerical models used in the wind energy forecasting process:

(1) Physical atmospheric models
(2) Statistical atmospheric models
(3) Wind plant output models and
(4) Forecast ensemble models

There are many types of models within each of these major categories. A particular forecast system may employ one or more of each type of model in its forecast procedure.

Physical atmospheric models are based upon the fundamental physical principles of conservation of mass, momentum and energy and the equation of state for air. These models are actually a type of computational fluid dynamic (CFD) model that has been specially adapted to simulate the atmosphere. These models consist of a set of differential equations that are numerically solved on a three-dimensional data grid which has a finite resolution (i.e. the spacing between grid cells). There are many types of these models. They are based on the same basic physical principles but they differ in how the grids are structured, how the equations are numerically solved and how sub-grid scale processes (i.e. things that happen on scales which are smaller than the grid cells) are represented.

Physics-based atmospheric models fall into two broad categories: prognostic and diagnostic. Prognostic models are formulated to use their equations to step forward from an initial state and make predictions of the future state of the atmosphere. It is necessary to specify an initial state to start this forecast process. An initial state consists of a value for each of the model's variables at each grid cell. This is produced by processing all of the raw atmospheric data from the various sensor systems that were described earlier in this section.

Statistical atmospheric models are simply statistical techniques used for atmospheric applications. They are "atmospheric" models in the sense that atmospheric data is used as input and the output is an atmospheric variable or quantity that is linked to an atmospheric variable (such as wind energy output). Statistical models operate by creating a set of empirical equations from a sample of predictor and predictand (the quantity to be predicted) data called a "training sample". The form of the equations is dependent on the type of model that is used. Typically, the equations have numerical coefficients that must be determined. A modeling procedure uses an optimization scheme to select the coefficient values that yield the best relationship between the predictors and the predictand. The meaning of "best" in this context depends upon what optimization criterion is employed. An example of optimization criteria is the lowest mean absolute error or the lowest mean squared error. Once the coefficients are determined from the training sample, the resulting equations can be used to produce a forecast by inserting the current values of the predictors and calculating the value of the predictand. There are an enormous number of statistical models available for this type of an application. The most popular ones for atmospheric science applications are multiple linear regression and neural networks.

Statistical models are used in a number of different ways in wind energy forecast systems. In one mode, they can be used to adjust the predictions from the physics-based models. This mode is commonly called Model Output Statistics (MOS). However, they also can be used to make predictions directly from measured data. For example, a time series of generation data can be used to train a statistical model and make predictions of the future generation.

Wind plant output models are the relationships between the meteorological variables at the wind plant site and the plant's energy output. They can be formulated as physical models, statistical models or a hybrid of both types of models. In a statistical approach, the parameters measured by sensors on the plant's meteorological towers typically serve as the predictors and the power generation is the predictand. The simplest plant output model is a relationship between the wind speed measured at a meteorological tower and the total plant output. The result is a plant-scale equivalent to the "power curve" of an individual turbine. This simple model can be extended for example, by developing a separate relationship for ranges of wind directions. This may be useful in accounting for the orientation of the turbine layout relative to the wind direction. For example, the power production may be different when the wind blows along a row of turbines than when it blows across a row.

Wind plant models can also be formulated as physical models. In this approach, the variations in wind flow within the wind plant, the interaction of the wind with the turbines and the effect of turbine wakes on other turbines are explicitly modeled. This requires detailed information about the layout of turbines in the plant, the properties (terrain, roughness etc.) of the earth's surface within the plant and information about the turbine specifications. The physical models have the advantage of being able to produce a power generation forecast without a training sample. They also can explicitly account for changes in the operating structure of a plant such as when turbines go out of service. They also, in theory, account for more of the details of the plant-scale variation in wind and its impact on power production. However, these models are typically much more complex than statistical models and require much detailed data about the plant which may not be easily available. Also, as with almost all physical models, there are likely to be systematic errors in the forecasts due to simplifying assumptions included in the model physics, limited resolution or the inaccuracies in the input data (e.g. inaccurate roughness or terrain etc.). In most applications it is necessary to use a statistical model to adjust the forecasts of a physical plant output model to remove these systematic errors.

The typical use of plant output models in the forecast process is to convert a wind speed prediction for one or more meteorological towers within the wind plant area into a power generation forecast for the plant. However, it is not necessary to have an explicit wind plant output model in a forecast system since it is possible to go directly from external predictors to a power output forecast through the use of an atmospheric statistical model.

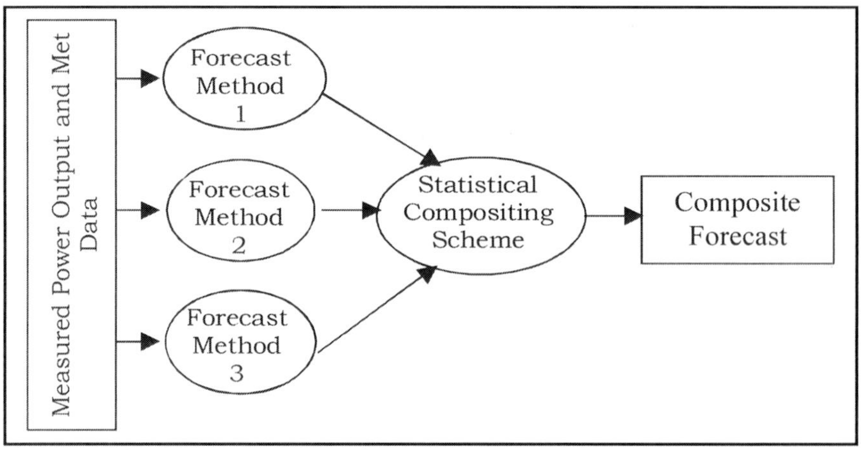

Fig.13.5. A schematic Depiction of the Ensemble Technique. This Arrangement Applies to both Short-term and Next-day Forecasts.

Forecast ensemble models are statistical models that produce an optimal forecast by compositing forecasts from a number of different techniques. The use of forecast ensemble models is based on research that has demonstrated that a composite of forecasts from an appropriate ensemble of forecast generating techniques is often superior to those produced by any one member of the ensemble. The method is schematically depicted in Fig 13.5. The fundamental concept is that if the errors in the forecasts produced by the different methods are unbiased and have a low degree of correlation with one another, the random errors from the individual forecasts will tend to offset each other, with the result that a composite of the forecasts will have a lower error than any individual forecast. If all of the input forecasts are highly correlated the impact of ensembling will be minimal. This means that the underlying forecast methods must be quite different in how they construct the relationships between the raw observational data and their forecasts or the type or amount of input data going into the methods must be significantly different. This "ensemble effect" is a well-known technique used by meteorologists in intermediate and extended range forecasting. The spread of the forecasts produced by the ensemble may also be related to the forecast uncertainty if the differences in the ensemble members are related to the primary factors that introduce uncertainty into the forecasts.

There are two fundamental strategies that can be used to generate an ensemble of forecasts. One strategy is to use the same forecast models and vary the input data within their range of uncertainty. The other is to use the same input data and to employ different forecast models or different configurations of the same model. The relative value of either strategy depends upon the sources of uncertainty in the forecast procedure. In practice, the sources of uncertainty vary with location, season and other factors, and thus the choice of the ensemble components and the number of members must be determined from experience and experimentation. This brief overview of forecast tools indicates that there is a large and diverse pool of tools that can be used to generate wind energy forecasts. The challenge is to use the optimal set of tools and configurations for a specific forecast application.

13.10 State-of-the-art: "Next-Hour" Forecasting

There are a wide variety of methods that have been or are being used to produce very short-term ("next-hour") wind energy generation forecasts. Fig. 13.6 provides a schematic depiction of the many components of the very short-term forecasting process and the ways they can be linked together to produce forecasts.

The simplest type of very short-term forecasts utilizes a time series of power generation data from the wind plant and a statistical procedure such as multiple linear regressions or a neural network to generate predictions of the future power output. These are often referred to as "persistence" or "autoregressive" models since their only source of information is the history of the plant's power output. These types of models can be enhanced through the use of a time series of meteorological data from the meteorological towers within the wind plant. The addition of this data can be handled in two ways. In the first approach the meteorological data is added to the pool of predictors and the power generation is predicted directly from the statistical model. In the second approach the meteorological data is used to forecast the meteorological inputs into a separate wind plant output model. The wind plant output model then uses these inputs to create an energy generation prediction.

The second approach may have an advantage if there is more than one met tower within the plant because it may be possible to capture some of the variability in meteorological conditions within the plant and hence produce a better energy generation forecast. Sophisticated statistical models such as neural networks may be able to find more subtle and complex relationships in the time series data and thereby generate better forecasts than simpler models such as linear regression. However, due to the fact that sophisticated statistical models usually have more adjustable parameters, they are more prone to "overfitting" problems if the training sample is not sufficiently large. Ultimately, all of these methods are limited by the fact that the input information is derived only from the history of conditions at the wind plant.

The next level of sophistication is to use multiple external data sources. The additional data sources can be used as input to the same types of statistical models used in the autoregressive approach. However, the number of predictors is larger. The additional data sources could include meteorological data from nearby meteorological towers or remote sensing systems. Another possibility is to use forecast output from a regional scale physical model. These models provide information about the larger scale trends in meteorological parameters but do not incorporate local area data and typically do not have the ability to resolve the local atmospheric and surface features that are critical to very short-term forecasting. However, some large-scale trends are well correlated with a local-scale response and hence the regional model data can at times add skill to the very short-term forecasts. A possibility that has yet to be thoroughly tested for very short-term wind power forecasting is to use a physical model with a high resolution grid to produce very short-term forecast simulations for the local-area surrounding the wind plant. In this approach, all of the available local-area data is assimilated into the initial state used to start the physical model simulation. This type of procedure has the potential to simulate the atmospheric features that cause the wind variations in the vicinity of the wind plant.

Another tool that can be used in the very short-term prediction process is a forecast ensemble model. As noted earlier, this is a statistical

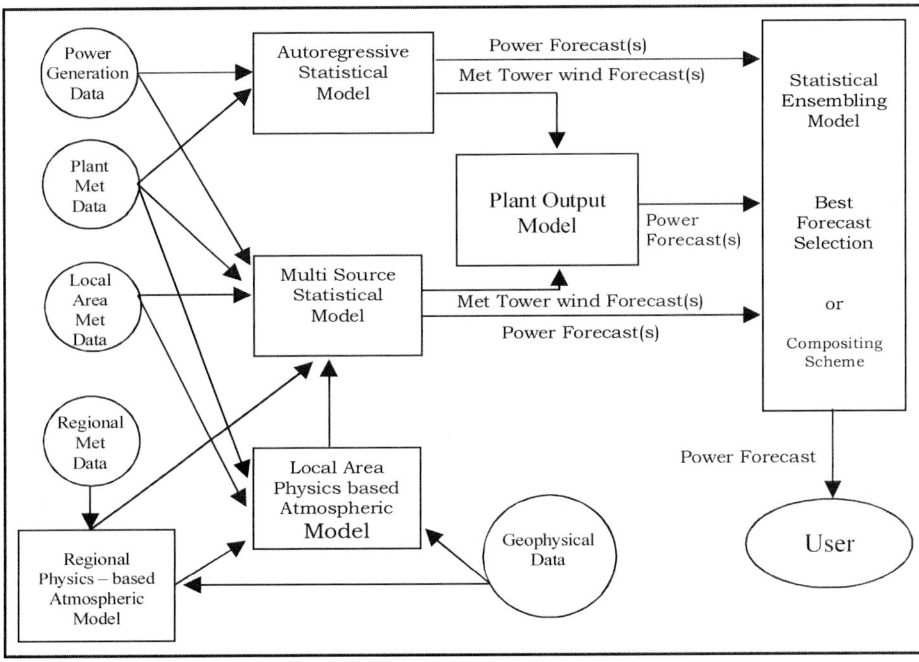

Fig 13.6. A Schematic Depiction of the Interrelationship of the Components of a Very Short-term Forecast System.

model that generates a composite forecast from a series of input forecasts generated by different forecast methods. After one completes an examination of the various methods that have been or could be used in the very short-term forecast process, the obvious questions are (1) what is the typical level of performance that can be expected from very short-term forecast methods? and (2) what is the variation in performance due to differences in methods, locations, seasons and weather regimes? Unfortunately, there has been no comprehensive controlled study of the performance of a broad spectrum of very short-term wind energy forecast methods over a diverse mix of atmospheric conditions. Most of the forecast performance evaluations have been done by forecast providers or researchers and not by an independent third party. It is difficult to draw conclusion from this array of evaluations because the evaluation methods, locations, and times are different.

13.11 State-of-the-art: "Day-Ahead" Forecasting

Short-term (day-ahead) forecast methods use essentially the same tools as very short-term forecast techniques. However, there are two important differences: (i) the importance of real-time data from the wind plant and its immediate environment is significantly reduced; and (ii) regional and sub-regional simulations with a physics-based atmospheric model play a much more significant role in the forecast process.

Almost all short-term forecast procedures begin with the grid point output from a regional scale physics-based atmospheric model. These models ingest data from a wide variety of sources over a large area and produce forecasts of regional scale weather system for a several day period. However, these models do not resolve the physical processes occurring in the local or mesoscale areas around individual wind plants. The three-dimensional output data from the regional-scale forecast simulations is the basic input into most intermediate-term wind energy forecast systems.

The forecast methods differ substantially from this point. Some forecast procedures attempt to go directly from the regional-scale forecast data to the local scale through the use of either diagnostic physical models, statistical models or a combination of both.

As in the case of very short-term forecast performance, a quantitative assessment of the state of the art in short-term wind energy generation forecast performance is difficult to obtain because most of the evaluations are done by individual forecast providers and researchers. The methods, locations and time periods used in these forecast performance evaluations vary substantially and therefore, it is difficult to determine the causes of differences in performance.

14

Improved Cookstoves

14.1 Introduction

Indoor air pollution causes significant health problems for the two billion people worldwide who rely on biomass fuels for their cooking and heating needs. Over the last 30 years awareness of the environmental and social costs of using traditional fuels and stoves has grown. At the same time, studies of the problem have resulted in proven strategies to reduce both fuel use and harmful emissions. Unfortunately, the local stoves currently available do not always represent the best designs that modern engineering can offer. Although open fires are often used wastefully, carefully operated open fires can be fuel efficient and clean burning when tested in the lab. In many situations, cooks are not overly concerned with fuel use, and studies have shown that when fuel is plentiful three-stone fires can use an excessive amount of wood to cook a small amount of food. But in other places where fuel is scarce, open fires can be carefully controlled so that fuel efficiency rivals many first generation improved cook stoves.

Modern cook stoves are designed to clean up combustion first. Then the hot gases can be forced to contact the pot and increasing efficiency without increasing harmful emissions. Fires can be clean burning when expert cooks push the sticks of wood into the fire as they burn, metering the fuel. The open fire can be a hot fire useful when food or drink needs to be prepared quickly. The energy goes into the pot, not into the cold body of a stove. The open fire can burn wood without making a lot of smoke; hot fires burn smoke as it is released from the wood (Fig.14.1). Unfortunately however, many fires used for cooking are built emphasizing simplicity of use and are wasteful and polluting. Modern stoves score higher when tested than even the most carefully operated fire in the laboratory.

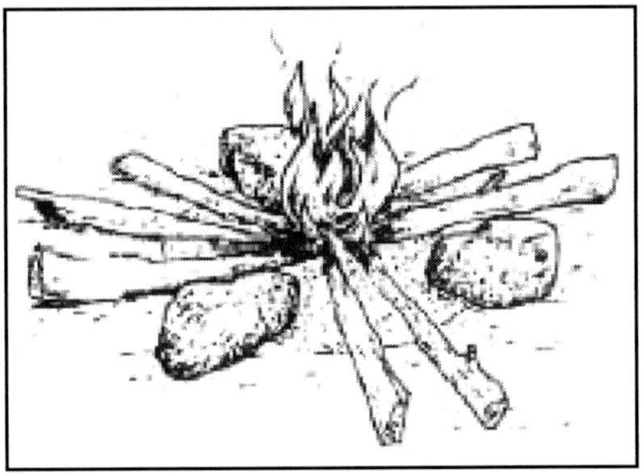

Fig 14.1. Traditional Firing

Good stoves can offer many advantages. Such stoves save wood and reduce smoke. Improved stoves can make cooking with fire easier, safer, faster, and can add to the beauty of the kitchen. A good stove is quicker to start, needs little tending, and can meet the specific needs of a cook.

Even an open fire is often 90 per cent efficient at the work of turning wood into energy. But only a small proportion, from 10 to 40 per cent, of the released energy makes it into the pot. Improving combustion efficiency does not appreciably help the stove to use less fuel. On the other hand, improving heat transfer efficiency to the pot makes a large difference.

Improving the combustion efficiency is necessary to reduce smoke and harmful emissions that damage health. Improving heat transfer efficiency can significantly reduce fuel use. Fire is naturally good for its job, but pots are not as good at capturing heat because they are inefficient heat exchangers. In order to reduce emissions and fuel use, the stove designer's job is to first clean up the fire and then force as much energy into the pot or griddle as possible. Both of these functions can be accomplished in a well engineered cooking stove. It is always best practice to add a chimney to any wood burning cooking or heating stove. Additionally, it is preferable to use a cleaner burning stove to protect air quality in and outside of the house. Chimneys that take smoke and other emissions out of the living space protect the family by reducing exposure to pollutants and health risks. Even cleaner burning stoves without a chimney can create unhealthy levels of indoor air pollution.

Unvented stoves should be used outdoors or in open areas. When chimneys are not affordable or practical using a hood over the fire, or opening windows, or making vents in the roof are all ways to decrease the levels of harmful pollution. The use of a cleaner burning stove can also be helpful in this regard. But however, all wood burning stoves should always be fitted with a functional chimney. Before designing of an improved cookstove, first, let's list the advantages of the three stone fire when compared to some stoves. No energy is absorbed into the mass of a stove body. High-mass stoves can absorb energy that could have gone into the pot. The three stone fire can boil water fairly quickly. Fire hits the bottom and sometimes the sides of the pot, exposing a lot of the pot to the hot gases. Sticks can be fed in at the appropriate rate as the tips burn, assisting complete combustion. A hot open fire can burn relatively cleanly. Every stove suffers because it has some mass that absorbs heat. But an improved stove can still achieve better combustion and fuel efficiency than an open fire.

14.2 Classification of Improved Cookstoves

(1) Classification on the basis of fuel used

(a) Firewood stoves

(b) Charcoal stoves

(c) Coal stove

(d) Lignite coke

(e) Sawdust stoves

(f) Rice husk stoves

(g) Bagasse stoves

(h) Dung cake stove

(i) Miscellaneous fuelled stoves.

(2) Classification on the basis of fixture

(a) Build-in type of stoves or fixed stoves

(b) Portable stoves

(3) Classification on the basis of shield
   (a) Open fire type stoves
   (b) Shielded stoves
   (c) Closed stoves

(4) Classification on the basis of material of construction used
   (a) Mud and clay stove
   (b) Concrete stoves
   (c) Mild steel stoves
   (d) Cast iron stove
   (e) Ceramic stoves
   (f) Stone stoves.

(5) Classification on the basis of pot opening
   (a) Single
   (b) Double pot
   (c) Multi pot.

(6) Classification on the basis of connective pot opening

(7) Classification on the basis of mounting of load.
   (a) Vessel mounted stoves
   (b) Close fit type stoves
   (c) Grooves stoves
   (d) Immersion system type stoves

(8) Classification according to accessories
   (a) Chimney type stoves
   (b) Damper type stoves
   (c) Baffle type stoves
   (d) Grate stoves
   (e) Lining stoves
   (f) Secondary air opening stoves
   (g) Blower stoves

(9) Classification on the basis of weight
   (a) Light weight stoves (1 to 3 kg)
   (b) Light medium stoves (3 to 10 kg)
   (c) Medium stoves (10 to 30 kg)
   (d) Medium heavy stoves (30 to 100 kg)
   (e) Heavy stoves (100 to 200 kg)
   (f) Very heavy stoves (more then 200 kg)

(10) Classification on the basis of stove life
   (a) Short
   (b) Medium
   (c) long (more than 10 years)

(11) Classification on the basis of average fuel burning rate

(12) Classification the basis of average power output

(13) Classification on the basis of average overall efficiency (Useful Heat/Input)

14.3 Improved Cookstove's Parts

Most of the stoves are composed of a few essential parts: combustion chamber, pot opening baffle, dampers, chimney etc. Particular combination of these results in efficient energy conversion. The effect of various parts of stove or stove parameters have got their own contribution in improving efficiency of the stoves. Some of these parameters are discussed as below-

(a) Height of combustion chamber

The distance between the floor and grate of stove and the bottom of the pot is one of the important parameters that affect the overall efficiency of energy conversion. If the height of the combustion chamber is too low, incomplete combustion with condensation of volatile, choking and smoke production results.

If the combustion chamber is too high, the flame mix with the surrounding sheets of air and lower the temperature, consequently lowering of convective and radiation components of the heat. Hence, the combustion chamber must have an optimum height for different burning rates.

(b) Pot openings

The pot openings can be radiant heat openings for the combustion chamber and convective heat openings for the subsequent ducts. Heat transfer between the hot gases of combustion and the pot depend, on openings for a particular vessel size and must be made as large as possible. Usually, pot openings are in the diameter range 180-300 mm for domestic operations. When spherical pots are used, the point of contact between the wall and the vessel is the maximum vessel diameter. The portion below this maximum diameter must be embedded into the stove chamber.

(c) Fuel openings

Energy stove needs an opening for feeding the fuel logs, but the opening causes convection and radiation losses. In addition, the fuel fed openings control the maximum burning rate by restricting fuel.

(d) Connecting ducts between the pots

The linear distance between the pots must be as little as possible i.e. 60 to 80 mm. However, a slope in the duct connecting the chamber and the chimney lowers the overall efficiency. It can be concluded that slopes must not be given for any flue duct, as these short circuits the hot gases and reduces efficiency.

(e) Air inlet parts for grated stoves

When the fuel feed door of grated stoves is closed to prevent air entry all the primary and secondary air to be drawn from the air inlet port by overcoming the resistance of the grate.

(f) Vessel mounting system

The vessel mounting system serves two functions. Firstly it holds the vessel load in a stable manner and secondly, it vents out the exhaust gases. In addition, there could be a combined system of close fit plus vessel immersion facilitated by a chimney.

14.4 How to Improve Combustion i.e. Make Less Harmful Pollution Compared to an Open fire

- Make sure there is good draft into the fire.
- Insulate around the fire to help it burn hotter. A hotter fire burns up more of the combustible gases and produces less smoke.
- Avoid using heavy, cold materials like earth and sand around the combustion chamber.
- Lift the burning sticks up off the ground so that air can scrape under the sticks and through the charcoal.
- Placing an insulated short chimney above the fire helps to increase draft and gives smoke, air, and fire a place to combine, reducing emissions.
- Meter the sticks of wood into the combustion chamber to make a hot, fierce, jumpy looking fire that does not make much charcoal. This type of fire will make less dangerous emissions, chimney clogging soot, and creosote. Heat only the burning part of the wood. Do not encourage the non-burning wood to make smoke.
- Limit the cold air entering the fire by using as small an opening as possible. Small openings into the fire also force the cook to use less wood, which can be burnt more efficiently.
- A certain amount of excess air is necessary for complete combustion. Preheating the air helps to maintain clean combustion.

14.5 How to Improve Fuel Efficiency i.e. get More Heat into the Pot

- Increase the temperature of the gas/flame contacting the pot, having the hot air scrapes against both the bottom and sides of the pot in a narrow channel, using a pot skirt.
- Increase the speed of the hot flue gases that scrape against the pot. The fast gases punch through a boundary layer of still air that keeps slower moving gases from scraping against the surface of the pot (or griddle.) Air is a poor heat transfer medium. It takes a lot of hot air to bring heat to the pot.
- Use metal rather than clay pots because metal conducts heat better than clay.

- The size of the fire determines the size of the channel gap in the pot skirt and the maximum efficiency of heat transfer. *Smaller fires that* can still please cooks but are not too big will be considerably more fuel efficient.

Use wide pots with large diameters. Using a wide pot creates more surface area to increase the transfer of heat. Make sure that the top of the stove slopes up toward the outer perimeter of the pot, as shown in Fig. 14.2.

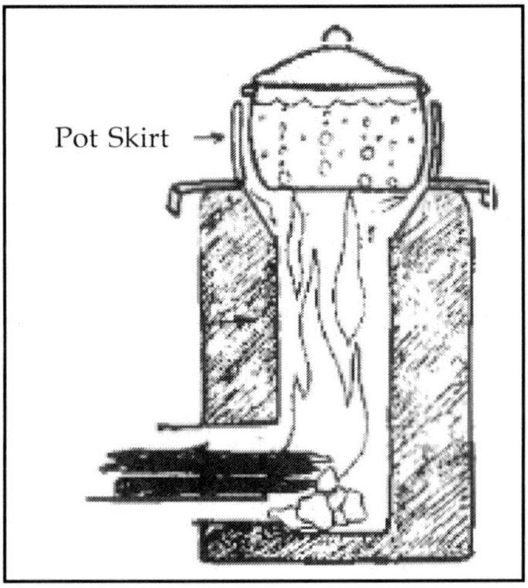

Fig 14.2. Appropriate Use of Pot Skirt

14.6 Ten Design Principles

14.6.1 Principle one

Whenever possible, insulate around the fire using lightweight, heat-resistant materials

If possible, do not use heavy materials like sand and clay; insulation should be light and full of small pockets of air. Natural examples of insulation include pumice rock, vermiculite, perlite, and wood ash. Light weight refractory brick (brick that has been fired and is resistant to cracking at high temperatures) can be made from locally available sources.

Insulation around the fire keeps it hot, which helps to reduce smoke and harmful emissions. Also, insulation around the fire keeps the heat from going into the stove body instead of into the pot. Unfortunately, metal does not last very long near a hot fire. However, locally made ceramic tiles can be found they are durable when used as walls for a combustion chamber. Loose insulation can surround this type of construction (Fig.14.3).

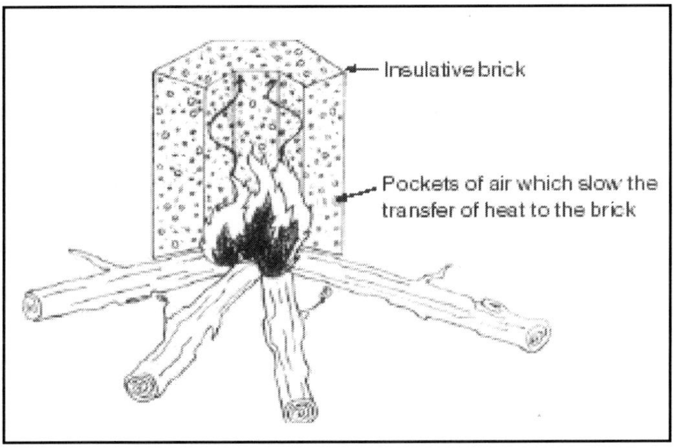

Fig 14.3 Insulation around the fire

14.6.2 Principal two
Place an insulated short chimney right above the fire

The combustion chamber chimney should be about three times taller than its diameter.

Placing a short chimney above the fire increases draft and helps the fire burn hot and fierce. Smoke will contact flame in the chimney become combust thus reducing emissions. Pots or surfaces to be heated are placed above the short chimney. A taller combustion chamber chimney, more than three times the width, will clean up more smoke, but a shorter chimney will bring hotter gases to the pot. The very tall combustion chamber chimney can develop too much draft bringing in too much cold air that will decrease heat transfer (Fig 14.4).

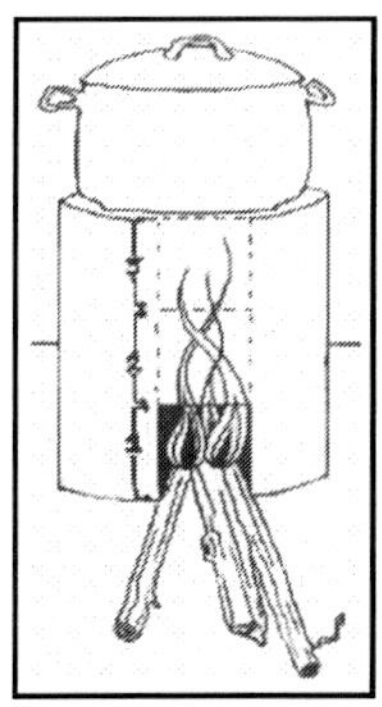

Fig. 14.4. An insulated short chimney above the fire

14.6.3 Principal three
Heat and burn the tips of the sticks as they enter the fire

If only the wood that is burning is hot there will be much less smoke. Try to keep the rest of the stick cold enough that it does not smolder and make smoke. The goal is to make the proper amount of gas so that it can be cleanly burned without making charcoal or smoke.

Smoke is un-burnt gas. It is harmful to breathe. Even cleaner looking combustion contains harmful emissions (Fig 14.5 a & b).

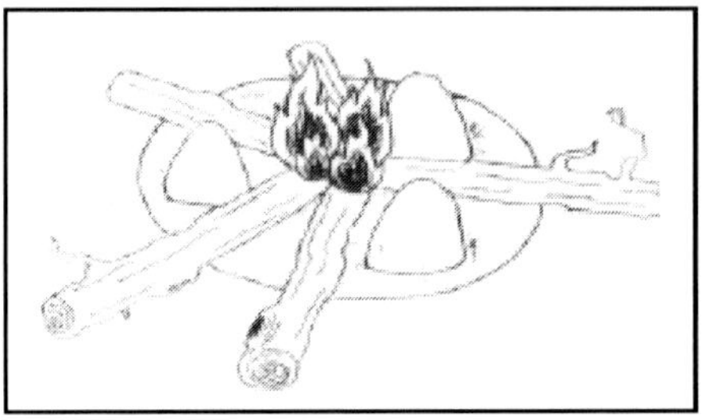

Fig 14.5 a. Cleaner Burning

Fig. 14.5 b. Smoldering Wood Makes Smoke

14.6.4 Principal four

How many sticks are pushed into the fire to create high and low heats

Adjust the amount of gas made and fire created to suit the cooking task. (Wood gets hot and releases gas. The gas catches fire and makes heat.) Fig. 14.6 a & b

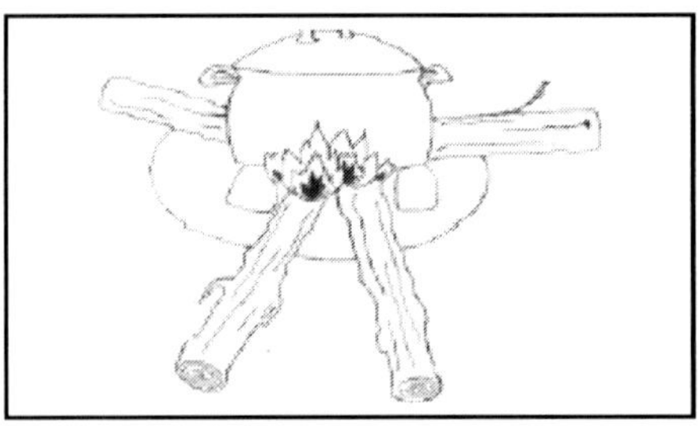

Fig 14.6a. Low Heat

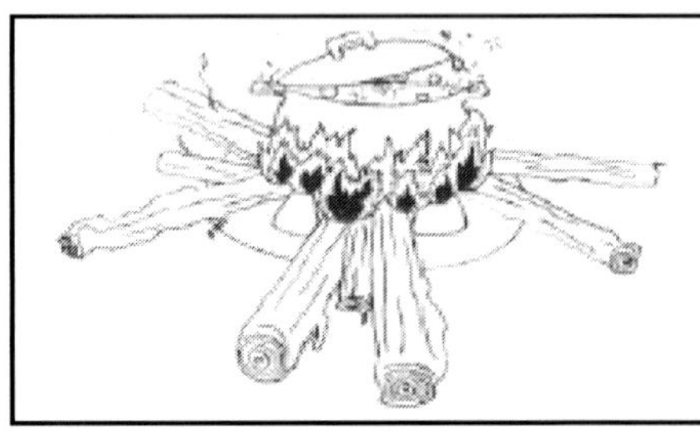

Fig 14.6b. High Heat

14.6.5 Principal five

Maintain a good fast draft through the burning fuel

Just as blowing on a fire and charcoal can make it hotter, having the proper amount of draft will help to keep high temperatures in stove. A hot fire is a clean fire (Fig 14.7).

14.6.6 Principal six

Too little draft being pulled into the fire will result in smoke and excess charcoal

But too much air just cools the fire and is not helpful. Smaller openings into the fire help to reduce excess air. Improving heat transfer to the pot or griddle is the most important factor that will reduce fuel use in a cooking stove. Improving combustion efficiency reduces pollution but is less important when trying to save firewood (Fig 14.8).

Fig. 14.7. Maintaining a Good Draft

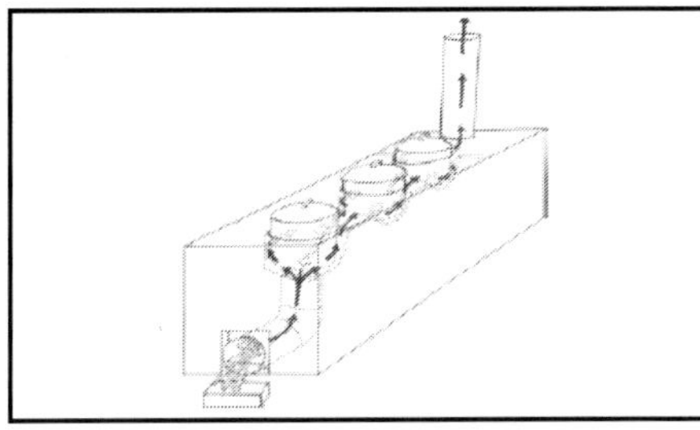

Fig. 14.8. Balancing the Air Flow in a Multipot Stove

14.6.7 Principal seven

The opening into the fire, the size of the spaces within the stove through which hot air flows and the chimney should all be about the same size. This is called maintaining constant cross sectional area, and helps to keep good draft throughout the stove. Good draft not only keeps the fire hot; it is also essential so that the hot air created by the fire can effectively transfer its heat into the pot. Air does not carry very much energy, so a lot of it needs to go through the stove in order to accomplish the task of heating food or water. The size of the openings is larger in more powerful stoves that burn more wood and make more heat. As a general rule, a door into the fire with a square opening of twelve centimeters per side and equally sized chimney and tunnels in the stove will result in a fire suited to family cooking.

14.6.8 Principal eight

Use a grate under the fire

Do not put the sticks on the floor of the combustion chamber. Air needs to pass under the burning sticks, up through the charcoal and into the fire. A shelf in the stove opening also lifts up sticks so air can pass underneath them. When burning sticks, it is best to have them close together and flat on the shelf, with an air space in between each stick. The burning sticks keep the fire hot, each fire reinforcing the other to burn more completely. It is optimum if the air passes under the shelf and through the coals so that when it reaches the fire it is preheated to help the gases for complete combustion. Air that passes above the sticks is not as helpful because it is colder and cools the fire. A hot raging fire is clean, but a cold fire can be very dirty (Fig 14.9).

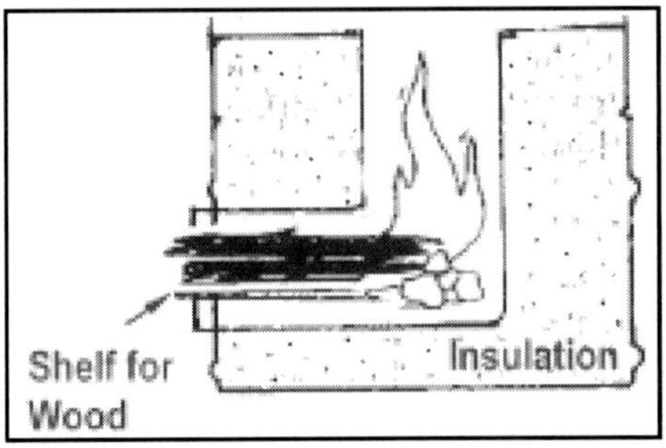

Fig. 14.9. Use of a Grate Under the Fire

14.6.9 Principal nine

Insulate the heat flow path

Cooks like stoves that boil water quickly. This can be especially important in the morning when family members need to get to work. If heat goes into the body of the stove, the pot boils less quickly. Why heat up fifty or one hundred kilograms of stove each morning when the desired result is to heat up a kilogram of food or a liter of water. Using insulating materials in the stove keeps the flue gases hot so that they can more effectively heat the pan or griddle. Insulation is full of air holes and is very light. Clay and sand or other dense materials are not insulating in nature. Dense materials soak up heat and divert it from cooking food.

14.6.10 Principal ten

Maximize heat transfer to the pot with properly sized gaps

Getting heat into pots or griddles is properly done with small channels. The hot flue gases from the fire are forced through these narrow channels, or gaps, where it is forced to scrape against the pot or griddle. If the gap is too large the hot flue gases mostly stay in the middle of the channel and do not pass their heat to the desired cooking surface. If the gaps are too small, the draft diminishes, causing the fire to be cooler, the emissions to go up and less heat to enter the pot. When designing a stove, it is possible to decrease the gap in the channel next to the pot or griddle until the fire becomes "lazy." Using trial and error, open up the gap little by little until the fire stays hot and vigorous.

14.7 Development of Improved Cookstove

The Department of Renewable Energy Sources, College of Technology and Engineering, MPUAT, Udaipur (Rajasthan) has developed two durable models of improved cookstoves made of bricks, sand & cement. These new models have been named as CHETAK single pot and UDAIRAJ double pot.

(a) Chetak Improved Cookstove

This improved cookstove is suitable for those who prefer one cooking at a time. The average thermal efficiency of this improved cookstove is 21.00 per cent and equally suitable for firewood, dung cakes and agro-waste.

(b) Udairaj Improved Cookstove

This is two pot improved cookstove specially designed for the rural users who want two cooking at a time. The average thermal efficiency of this improved cookstove is 26.00 per cent.

14.7.1 Traditional Practices

The traditional chulha made of three stone style and of mud is generally used by villagers. Mostly villagers are cooking their food on traditional chulha, which is installed by beneficiaries by placing three stones. There is no provision for smoke emission in this traditional chulha.

System Details

(i) Chetak Improved Cookstoves

The overall size of this improved cookstove is 60 x 45 x 24 cm. This model is suitable for those who prefer one cooking at a time. The average thermal efficiency of this improved cookstove is 21.00 per cent.

This Chulha is suitable for firewood, agro waste, and dung cakes. Bigger sized wood can also be used in this chulha at the fuel burning rate of 1.00 kg/hr. The passage between the firebox chamber and the inlet of the chimney pipe is horizontal, which flows smoky gases outside the kitchen.

A mixture of cement and sand is used to fix up the bricks. After this a thin plaster is made on the outer surface of the chulha. The centre also developed a metallic mould for Chetak Chulha for speedy and accurate construction in the field. The specification of the Chetak chulha is as below:

1. Body material brick masonry with cement plaster
2. Major dimensions

(a) Body : 600 mm x 450 mm x 240 mm

(b) Fire box : Diameter 240 mm

(c) Fire box opening : Trapezoidal in shape,

Size-240 mm x 120 mm x 240 mm

(ii) Udairaj-Double Pot Chulha

According to feedback received from field that the model should be bigger than the previous double pot chulha models. Medium sized double pot Improved Chulha with higher thermal efficiency as compared to earlier model is also developed.

The "Udairaj" model is specially designed to meet the cooking requirement of the family of 6-12 persons and is of permanent nature. The special feature of this improved chulha is bigger size of fire box, which is suitable for big size firewood and light agro-waste type of fuel.

The diameter of the 1st pot is 24 cm, which is suitable for households for baking of "Chapatti". The diameter of 2nd pot is 20 cm and suitable for cooking vegetables and boiling of milk etc. The burning of wood is very efficient without back fire, also CO/CO_2 ratio is within permissible limit.

This improved chulha works with an average thermal efficiency of 26 per cent at the fuel burning rate of 1.0 kg/hr. A mixture of cement and sand is used to fix up the bricks. After this plaster is made on the outer surface of the chulha.

A metallic mould set has also been developed for quick and accurate installation of this improved chulha in the field. The Specification of the Udairaj chulha is as follows:

1. Body material : Brick masonry with cement plaster
2. Major dimensions :

   (a) Body 850 mm x 400 mm x 250 mm

   (b) Fire box first pot dia. 240 mm & second pot dia. 200 mm.

   (c) Fire box opening 240 mm x 160 mm

   (d) Tunnel length Diameter 2.5", first pot to second pot-100 mm length second pot to chimney-80 mm length

Details of the Accessories

The improved cookstoves namely Chetak & Udairaj are made of cement brick masonry. Essentially it require a firebox made of brick cum cement mortar, plastered with cement. Similarly a AC pipe of 10 feet long and 3 inch diameter is also required for creating draft and providing sufficient secondary air for proper combustion. In case of double pot Udairaj chulha, a connecting tunnel is also provided for flowing fire from first pot to second pot.

Performance Data

The performance of Chetak (Single-pot) and Udairaj (Double-pot) is carried out by standard water boiling test. The sun dried wood *Acacia nilotica* (Desi babool) is used for testing and having calorific value of 3700 Kcal/kg. This is used @1kg/hr. The thermal efficiency of Chetak and Udairaj work out to 21.34 and 25.54 per cent respectively.

Cost of the Technology

The Chetak and Udairaj improved cookstove is made of material of permanent nature and having at least five years useful life.

Estimate for Construction of "Chetak" single pot durable improved cook stove and "Udairaj" double pot durable improved cook stove is given in Table 14.1 & Table 14.2 respectively.

Table 14.1. Cost Estimates of Chetak Single Pot Cookstove.

S. No.	Material	Quantity	Cost (Rs.)
1.	Bricks	20 no.	35
2.	Cement	8 kg	30
3.	Sand	30 kg	10
4.	3″ dia A.C. Chimney Pipe-10′ length, cowl, tunnel (2″ dia, 4″ length)	1 set	80
5.	Mason cost		45
		TOTAL	**200**

Table 14.2: Cost Estimates of Udairaj Double Pot Cookstove.

S. No.	Material	Quantity	Cost (Rs.)
1.	Bricks	30 no.	45
2.	Cement	12 kg	35
3.	Sand	50 kg	10
4.	3″ dia A. C. Chimney Pipe-10′ length, cowl, tunnel (2½″ dia , 8″ length)	1 set	80
5.	Mason cost		60
		Total	**230**

The single-pot cooking stove is simple to construct and maintain and propagated widely in the tribal belt of Rajasthan. It is also suitable for other families who prefer single-pot cooking. The two-pot stove, although it needs more maintenance to keep it properly functional, provides effective fuel saving and reduces the level of pollution in the kitchen. The stove has been propagated and accepted in rural areas of many states in India.

Construction material for Improved Cookstoves

The construction technique used in improved chulha should be simple with low demand of material, low in cost and easy to build and beneficiaries should easily know it. The various material used are given in Table- 14.3 as follow-

Table 14.3: Materials of Construction for Improved Cookstove

S. No.	Material	Quality Needed
1.	Bricks	Good quality bricks preferably machine made first class bricks with uniform size and shape
2.	Cement	Pure portland with no impurities, packed in polythene bags.
3.	Sand	Fine and coarse with good quality.
4.	Asbestos-cement pipes for removal of smoke	Good quality with no leakage and should have uniform diameter and length as required.

Safety Requirement

At the time of construction of these chulha care must be taken such that the internal skeleton made of bricks and cement mortar be made within one day and then it must be cure for one more day after that cement plaster should be made. It is also recommended that proper curing of these chulha be made for atleast 15 days, thereafter it may be put in use. It has also been observed that the cement plaster may get cracked at excess temperature, therefore it is recommended that when cookstoves put in use, gradually for 2 to 3 days it may be exposed to heat through small quantity of wood burning and thereafter it may be put in regular use.

During cooking on improved cookstoves in double pot both pot should be used simultaneously. Regularly cleaning of chimney pipe & connecting tunnels is required for smoke emission.

Trouble shooting

The following troubles are generally observed in the proper operation of improved cookstoves.

(a) The sufficient fire is not available in the second pot in case of Udairaj. For this required quantity of fuel i.e. at the rate of 1.00 kg/hr. is to be burnt.

(b) The outer skirt of the chulha is also heated up due to continuous use of chulha. Therefore it is recommended that beneficiary using those chulha should not seat closed to the body of chulha.

(c) Many times sufficient draft is not created, it is because of carbon deposit on the inner surface of chimney (pipe).

(d) Because of poor quality of cement use, there are chances of breakage of cement plaster on the surface. Hence, perfect check must be made at the time on the quality of material during construction of chulha.

(e) Cement may develop cracks at excessive temperature, if not properly cured with water. Most of cement requires at least 15 days proper curing before use.

(f) Good quality bricks be used.

14.8 Durable Improved Biomass Cookstove for Agro-Industrial and Community Applications

A very large variety of combustion devices and biomass stoves are being used for community cooking and cottage/small agro-industrial applications. Most of these stoves do not have adequate provision for smoke emission. Their thermal efficiency normally varies between 10-20 per cent and the persons working in the kitchen or the industry are exposed to very high level of pollution. Single and double pot cook stoves may be used for community and cottage/small agro-industrial applications. Both the models are made using bricks, sand, cement, A.C. pipe and steel grate. The single pot stove, which has a power rating of around 10 KW, is suitable for meeting process heat requirements at industrial level and has thermal efficiency of around 35 per cent. The double pot improved cook stove is primarily designed for community applications, hotels and roadside restaurants/*dhabas*. This has a power rating of around 5.5 kW and a thermal efficiency of more than 40%. Fuel wood, densified briquettes, woody crop residues and dung cakes may be used as fuel in these stoves. Single pot stove has been designed for supplying process heat of around 20,000 kcal/h and is suitable for use of firewood, biomass briquettes, dung cakes and woody crop residues. Double pot cook stove has been designed to cook food for 50-60 persons. The main burner is used for boiling, baking an frying operations while the 2nd burner is used for heating/boiling operations. It has been provided with bigger size firebox, and therefore, can burn big size firewood and also woody crop residues. Major specifications of the stoves are given in Table 14.4.

Table 14.4 Major Specific of Single and Double Pot Stove.

S.No.	Description	Single Pot	Double Pot
1.	Body size, mm	500 (diameter)	1775 x 1100 x 450
2.	Firebox diameter		
	- 1st pot, mm	300	450
	- 2nd pot, mm	—	450
3.	Firebox opening		
	- shape	Rectangular	Rectangular
	- size, mm	300 x 300	300 x 250
4.	Power rating, kW	10.0	5.5

The improved cook stoves namely single pot industrial improved cook stove and double pot community improved cook stove are made of cement-brick masonry. Essentially it requires a firebox made of bricks and cement mortar. An AC pipe of 10 cm diameter and 3.3 m in length is provided for creating sufficient suction for exhaust removal and providing adequate secondary air for proper combustion of the fuel. In case of the double pot community cook stove, a connecting tunnel is also provided for flowing fire from the first pot to the second pot. Specifications of the materials recommended for construction are given in Table 14.5.

Table 14.5 Specifications of the Materials Used for Construction

S.No.	Material	Important Specifications
1.	Bricks	Good quality first cass bricks with uniform size and shape preferably machine made
2.	Cement	Pure Portland cement with no impurities, packed in polythene bags
3.	Sand	Fine and Coarse with good quality
4.	Asbestos-cement pipes for removal of smoke	Good quality with no cracks and should have uniform diameter and length as required.

Performance

Performance of both the designs based on standard water boiling test is given in Table 14.6. Result are based on Sun-dried desi babool (*Acacia nilotica*) having average calorific value of 3700 kcal/kg as fuel. Average thermal efficiency of the single pot stove for 8 kg/h burniong rate was found to be 34.8 per cent and for the two pot community cook stove for a fuel burning rate of 5.5 kg/h was found to be 41.5 per cent.

Performance data collected reveal an average fuel wood consumption of 9.5 kg/h for the improved cook stove as compared to 11.3 kg/h for the existing stove. This indicates a saving of 15.9 per cent in fuel wood consumption over the existing stove or saving of about 14.4 kg fuel wood per day. The performance of double pot community cook stove has been evaluated and observed that average fuel wood consumption was 3.4 kg/h as against 4.4 kg/h for the existing device. This represented a saving of about 23 per cent in fuel wood consumption or about 8.4 kg fuel wood per day.

Table 14.6 Thermal Efficiency of Single Pot Industrial and Double Pots Durable Improved Cook Stoves with Desi Babool (*Acacia nilotica*)

Model/test replication	Quantity of Water Filled	Avg. fuel burning Evaporated	Avg. power rating kw rate, kg/h	Thermal efficiency, %
Single				
Pot	50.0	5.4	8.00	9.87
35.45				
- First	0	5.2	8.00	9.92
34.24				
- Second	50.0	5.3	8.00	9.90
34.85				
Mean	0			
	50.0			
	0			
Double				
Pot	25.0	22.26	3.50	5.46
42.07				
1st Pot 1	0	20.60		
Pot 2	23.0	22.27	3.50	5.50
41.01				
2nd Pot 1	0	20.65		
Pot 2	25.0	—	3.50	5.48
41.54				
Mean	0			
	23.0			
	0			
	—			

Construction and Maintenance

The improved cook stoves may be got constructed through any skilled mason as per drawing details given in this bulletin and using the materials of appropriate quality. The internal skeleton should be made using good quality, first class bricks laid with cement mortar and cured for at least one day before carrying out plastering with sand cement mortar. Asbestos cement pipe of 76 mm diameter and 3.3 m length should be fixed properly at the stove end and made waterproof at the ceiling/roof for trouble free operation. One tunnel of 50 mm diameter, 100 mm length and 2 tunnels of 60 mm diameter and 200 mm length are used for single pot and double pot stoves, respectively. The skeleton may be plastered with 12 mm thick cement-sand mortar and cured thoroughly for 15 days to avoid cracking of the plaster. Fuel burning may be gradually increased to full scale over a period of 3 days. In case of double pot stove, both the stoves should be used simultaneously. The outer surface of these stoves gets heated up due to continuous use. A mud plaster may be applied on the outer surface to avoid skin burns.

Table 14.7 Details of Materials required and Cost Estimates for Construction of Single Pot Industrial Improved Cook Stove and Double Pot Community Improved Cook Stove.

S.No.	Item	Single Pot		Double Pot	
		Quantity	Cost, Rs.	Quantity	Cost, Rs.
1.	Standard size bricks	200 Nos.	500	350 Nos.	875
2.	Cement	75 kg	225	100 kg	300
3.	Fine and coarse sand	200 kg	100	400 kg	200
4.	A.C. pipe 76 mm and 3.3 m long, cowl, tunnel	1 set	150	1 set	150
		—	200	—	250
5.	Mason and labour cost				
		Total	1175		1175

Cost of construction

Details of materials used and estimate of costs for single pot industrial and two pot community stoves are given in Table 14.7. Approximate cost of construction has been estimated to be Rs. 1175 for the single pot industrial stove and Rs. 1775 for the double pots community stove. The estimated service life of both the stoves is 5 years.

15

Fuel Cell Technology

15.1 Fuel Cell

Fuel cell is a electrochemical device that convert chemical energy of a reaction directly into electrical energy. The basic physical structure of a fuel cell consists of an electrolyte layer in contact with a porous anode and cathode on either side. A schematic representation of a fuel cell with the reactant or product gases and the ion conduction flow directions through the cell is shown in Fig. 15.1

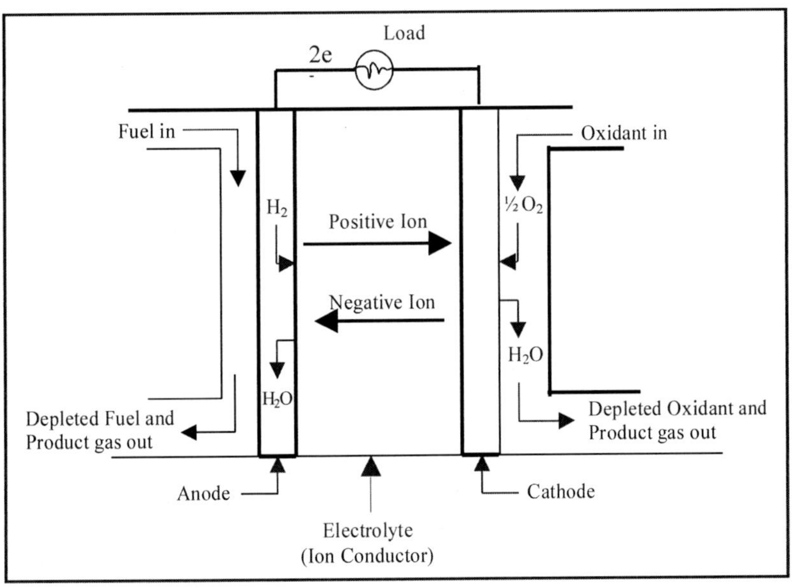

Fig 15.1. Schematic of a Fuel Cell

In a typical fuel cell, gaseous fuels are fed continuously to the anode (negative electrode) compartment and an oxidant (i.e., oxygencompartment, the electrochemical reactions take place at the electrodes to produce an electric current. A fuel cell, although having components and characteristics similar to those of a typical battery, differs in several respects. The battery is an energy storage device. The maximum energy available is determined by the amount of chemical reactant stored within the battery itself. The battery will cease to produce electrical energy when the chemical reactants are consumed (i.e., discharged). In a secondary battery, the reactants are regenerated by recharging, which involves putting energy into the battery from an external source. The fuel cell, on the other hand is an energy conversion device that theoretically has the capability of producing electrical energy for a period till the fuel and oxidant are supplied to the electrodes. In reality degradation primarily corrosion or malfunction of components limits the practical operating life of fuel cells.

The ion can be either a positive or a negative charge carrier (surplus or deficit of electrons). The fuel or oxidant gases flow past the surface of the anode or cathode opposite the electrolyte and generate electrical energy by the electro chemical oxidation of fuel, usually hydrogen, and the electrochemical reduction of the oxidant, usually oxygen.

Gaseous hydrogen is the fuel of choice for most applications in fuel cell, because of following reasons:-

(1) High reactivity when suitable catalysts are used.

(2) Its ability to be produced from hydrocarbons for terrestrial applications.

(3) Its high energy density when stored cryogenically for closed environment applications, such as in space.

Similarly, the most common oxidant is gaseous oxygen, which is readily and economically available from air for terrestrial applications and again easily stored in a closed environment. A three-phase interface is established among the reactants, electrolyte and catalyst in the region of the porous electrode. The nature of this interface play a critical role in the electrochemical performance of a fuel cell, particularly in those fuel cells with liquid electrolytes. In such fuel cells, the reactant gases diffuse through a thin electrolyte film that wets portions of the porous electrode and react electro chemically on their respective electrode surface.

If the porous electrode contains an excessive amount of electrolyte, the electrode may "flood" and restrict the transport of gaseous species in the electrolyte phase to the reaction sites. The consequence is a reduction in the electro chemical performance of the porous electrode. Thus, a delicate balance must be maintained among the electrode, electrolyte and gaseous phases in the porous electrode structure. Much of the recent effort in the development of fuel cell technology has been devoted to reducing the thickness of cell components while refining and improving the electrode structure and the electrolyte phase, with the aim of obtaining a higher and more stable electrochemical performance while lowering cost.

The electrolyte not only transports dissolved reactants to the electrode, but also conducts ionic charge between the electrodes and thereby completes the cell electric circuit, as illustrated in Fig. 17.1. It also provides a physical barrier to prevent the fuel and oxidant gas streams from directly mixing.

The functions of porous electrodes in fuel cells are:

1. To provide a surface site where gas/liquid ionization or de-ionization reactions can take place.
2. To conduct ions away from or into the three phase interface once they are formed (so an electrode must be made of materials having good electrical conductance), and
3. To provide a physical barrier that separates the bulk gas phase and the electrolyte.

In order to increase the rates of reactions, the electrode material should be catalytic as well as conductive, porous rather than solid. The catalytic function of electrodes is more important in lower temperature fuel cells and less so in high temperature fuel cells because ionization reaction rates increase with temperature. It is also a corollary that the porous electrodes must be permeable to both electrolyte and gases, but not such that the media can be easily "flooded" by the electrolyte or "dried" by the gases in a one-sided manner.

A variety of fuel cells are in different stages of development. They can be classified depending on the combination of type of fuel and oxidant whether the fuel is processed outside (external reforming) or inside (internal reforming) the fuel cell, the type of electrolyte, the temperature of operation, whether the reactants are fed to the cell by internal or external manifolds, etc.

15.2 Classification of Fuel Cells

The most common classification of fuel cells is based on the type of electrolyte used and are classified as.

(a) Polymer Electrolyte Fuel Cell (PEFC)

(b) Alkaline Fuel Cell (AFC)

(c) Phosphoric Acid Fuel Cell (PAFC)

(d) Molten Carbonate Fuel Cell (MCFC)

(e) Intermediate Temperature Solid Oxide Fuel Cell (ITSOFC) and

(f) Tubular Solid Oxide Fuel Cell (TSOFC).

The operating temperature of these fuel cells are ranging from ~80°C for PEFC, ~100°C for AFC, ~200°C for PAFC, ~650°C for MCFC, ~800°C for ITSOFC, and 1000°C for TSOFC. The operating temperature and useful life of a fuel cell depend on the physico chemical and thermo mechanical properties of materials used in the cell components (i.e., electrodes, electrolyte, interconnect, current collector, etc.). Aqueous electrolytes are limited to temperatures of about 200°C or lower because of their high water vapor pressure and/or rapid degradation at higher temperatures. The operating temperature also depends on the type of fuel that can be used in a fuel cell. The low-temperature fuel cells with aqueous electrolytes are in most practical application, restricted to hydrogen as a fuel. In high-temperature fuel cells CO and even CH_4 can be used because of the inherently rapid electrode kinetics and the lesser need for high catalytic activity at high temperature. The higher temperature cells can favor the conversion of CO and CH_4 to hydrogen, then use the equivalent hydrogen as the actual fuel.

(a) Polymer Electrolyte Fuel Cell (PEFC)

The electrolyte in this fuel cell is an ion exchange membrane (fluorinated sulfonic acid polymer or other similar polymer) that is an excellent proton conductor. The only liquid in this fuel cell is water thus corrosion problems are minimal. Water management in the membrane is critical for efficient performance of the fuel cell. It must operate under conditions where the byproduct water does not evaporate faster than it is producedbecause the membrane must be hydrated. Because of the limitation on the operating temperature imposed by the polymer, usually less than 120°C and because of problems with water balance, a H_2-rich gas with minimal or no CO is used. Higher catalyst loading (Pt in most cases) than that used in PAFCs is required for both the anode and cathode.

(b) Alkaline Fuel Cell (AFC)

The electrolyte in this fuel cell is concentrated (85 wt%) KOH in fuel cells operated at high temperature (~250°C) or less concentrated (35-50 wt%) KOH for lower temperature (<120°C) operation. The electrolyte is retained in a matrix (usually asbestos) and a wide range of electrocatalysts can be used (e.g., Ni, Ag, Metal Oxides, Spinels, and Noble metals). The fuel supply is limited to non-reactive constituents except for hydrogen. CO is a poison and CO_2 will react with the KOH to form K_2CO_3, thus altering the electrolyte. Even the small amount of CO_2 in air must be considered with the alkaline cell.

(c) Phosphoric Acid Fuel Cell (PAFC)

Phosphoric acid concentrated to 100 per cent is used for the electrolyte in this fuel cell which operates at 150 to 220°C. At lower temperatures phosphoric acid is a poor ionic conductor and CO poisoning of the Pt electrocatalyst in the anode becomes severe. The relative stability of concentrated phosphoric acid is high compared to other common acids, consequently the PAFC is capable of operating at the high end of the acid temperature range (100 to 220°C). In addition, the use of concentrated acid (100 per cent) minimizes the water vapor pressure so water management in the cell is not difficult.

(d) Molten Carbonate Fuel Cell (MCFC)

The electrolyte in this fuel cell is usually a combination of alkali carbonates which is retained in a ceramic matrix of $LiAlO_2$. The fuel cell operates at 600 to 700°C, where the alkali carbonates form a highly conductive molten salt with carbonate ions providing ionic conduction. At the high operating temperatures in MCFC, Ni (anode) and nickel oxide (cathode) are adequate to promote reaction. Noble metals are not required.

(e) Intermediate Temperature Solid Oxide Fuel Cell (ITSOFC)

The electrolyte and electrode materials in this fuel cell are basically the same as used in the TSOFC. The ITSOFC operates at a lower temperature, typically between 600 to 800°C. For this reason, thin film technology is being developed to promote ionic conduction, alternative electrolyte materials are also being developed.

(f) Tubular Solid Oxide Fuel Cell (TSOFC)

The electrolyte in this fuel cell is a solid, nonporous metal oxide, usually Y_2O_3-stabilized ZrO_2. The cell operates at 1000°C where ionic conduction by oxygen ions takes place. Typically, the anode is Co-ZrO_2 or Ni-ZrO_2 cermet, and the cathode is Sr-doped $LaMnO_3$.

In low-temperature fuel cells (PEFC, AFC, PAFC) protons or hydroxyl ions are the major charge carriers in the electrolyte, whereas in the high-temperature fuel cells, MCFC, ITSOFC and TSOFC carbonate ions and oxygen ions are the charge carriers, respectively.

Even though the electrolyte has become the predominant means of characterizing a cell, another important distinction is the method used to produce hydrogen for the cell reaction. Hydrogen can be reformed from natural gas and steam in the presence of a catalyst starting at a temperature of ~760°C. The reaction is endothermic. MCFC, ITSOFC and TSOFC operating temperatures are high enough that reforming reactions can occur within the cell, a process referred to as internal reforming. The reforming reaction is driven by the decrease in hydrogen as the cell produces power. This internal reforming can be beneficial to system efficiency because there is an effective transfer of heat from the exothermic cell reaction to satisfy the endothermic reforming reaction. A reforming catalyst is needed adjacent to the anode gas chamber for the reaction to occur. The cost of an external reformer is eliminated and system efficiency is improved, but at the expense of a more complex cell configuration and increased maintenance issues of the reaction sites. Porous electrodes used in fuel cells, achieve much higher current densities. These high current densities are possible because the electrode has a high surface area, relative to the geometric plate area that significantly increases the number of reaction sites and the optimized electrode structure has favorable mass transport properties. In an idealized porous gas fuel cell electrode, high current densities at reasonable polarization are obtained when the liquid (electrolyte) layer on the electrode surface is sufficiently thin so that it does not significantly impede the transport of reactants to the electro active sites and a stable three-phase (gas/ electrolyte/electrode surface) interface is established. When an excessive amount of electrolyte is present in the porous electrode structure, the electrode is considered to be "flooded" and the concentration polarization increases to a large value.

15.3 Advantages and Disadvantages

The fuel cell differs in their materials of construction, fabrication techniques and system requirements etc. These distinctions result in individual advantages and disadvantages that decide the potential of the various cells for different applications.

(i) PEFC

The PEFC, like the SOFC, has a solid electrolyte. As a result, this cell exhibits excellent resistance to gas crossover. In contrast to the SOFC, the cell operates at a low temperature (80°C). This result in a capability to bring the cell to its operating temperature quickly, but the rejected heat cannot be used forcogeneration or additional power. This type of cell can operate at very high current densities compared to the other cells. However, heat and water management issues may limit the operating power density of a practical system. The PEFC tolerance for CO is in the low ppm level.

(ii) AFC

The AFC was one of the first modern fuel cells developed in 1960. The application at that time was to provide on-board electric power for the Apollo space vehicle. Desirable attributes of the AFC include its excellent performance on hydrogen (H_2) and oxygen (O_2) compared to other fuel cells due to its active O_2 electrode kinetics and its flexibility to use a wide range of electro catalysts, an attribute that provides development flexibility. When development was in progress for space application, terrestrial applications were also investigated for this type of cell. It was investigated that pure hydrogen would be required in the fuel stream because CO_2 in any reformed fuel reacts with the KOH electrolyte to form a carbonate reducing the electrolyte's ion mobility. Pure H_2 could be supplied to the anode by passing a reformed, H_2-rich fuel stream by a precious metal (palladium/silver) membrane. The H_2 molecule is able to pass through the membrane by absorption and mass transfer and into the fuel cell anode. However, a significant pressure differential is required across the membrane and the membrane is prohibitive in cost. Even the small amount of CO_2 in ambient air the source of O_2 for the reaction, would have to be scrubbed.

(iii) PAFC

The CO_2 in the reformed fuel gas stream and the air does not react with the electrolyte in a phosphoric acid electrolyte cell, but is a diluent. This attribute and the relatively low temperature of the PAFC made it a prime, early choice for terrestrial application. Although its performance is somewhat lower than the alkaline cell because of the cathode's slow oxygen reaction rate and although the cell still requires hydrocarbon fuels to be reformed into an H_2-rich gas the PAFC system efficiency improved because of its higher temperature environment and less complex fuel conversion (no membrane and attendant pressure drop). There is no need to scrub CO_2 from the process air. The rejected heat from the cell is high enough in temperature to heat water or air in a system operating at atmospheric pressure. Some steam is available in PAFCs, a key point in expanding cogeneration applications.

PAFC systems achieve about 37 to 42 per cent electrical efficiency (based on the LHV of natural gas). This is at the low end of the efficiency goal for fuel cell power plants. PAFCs use high cost precious metal catalysts such as platinum. The fuel has to be reformed external to the cell and CO has to be shifted by a water gas reaction to below 3 to 5 vol per cent at the inlet to the fuel cell anode or it will affect the catalyst. These limitations have prompted development of the alternate, higher temperature cells, MCFC and SOFC.

(iv) MCFC

Many of the disadvantages of the lower temperature as well as higher temperature cells can be alleviated with the higher operating temperature MCFC (approximately 650°C). This temperature level results in several benefits like

(1) The cell can be made of commonly available sheet metals that can be stamped for less costly fabrication.

(2) The cell reactions occur with nickel catalysts rather than with expensive precious metal catalysts.

(3) Reforming can take place within the cell provided a reforming catalyst is added (results in a large efficiency gain).

(4) CO is a directly usable fuel and the rejected heat is of sufficiently high temperature to drive a gas turbine and produce a high pressure steam for use in a steam turbine or for cogeneration.

(5) It operates efficiently with CO_2-containing fuels such as bio-fuel derived gases. This benefit is derived from the cathode performance enhancement resulting from CO_2 enrichment.

The MCFC has some disadvantages, like the electrolyte is very corrosive, mobile and a source of CO_2 is required at the cathode (usually recycled from anode exhaust) to form the carbonate ion. Sulfur tolerance is controlled by the reforming catalyst and is low which is the same for the reforming catalyst in all cells. Operation requires use of stainless steel as the cell hardware material. The higher temperatures promote material problems, particularly mechanical stability that impacts life.

(v) ITSOFC

The intermediate temperature solid oxide fuel cell combines the best available attributes of fuel cell technology development with intermediate temperature (600-800°C) operation. Ceramic components are used for electrodes and electrolytes, carbon doesnot deposit on these ceramic materials therefore this fuel cell may accept hydrocarbons and carbon monoxide in the fuel. Internal reforming is practically at temperatures above 650°C. Moreover, use of solid state components avoids design issues such as corrosion and handling, inherent in liquid electrolyte fuel cells. The reduced temperature from the TSOFC allows stainless steel construction, which represents reduced manufacturing costs over more exotic metals. The disadvantages of ITSOFCs are that electrolyte conductivity and electrode kinetics drop significantly with lowered temperature.

(vi) TSOFC

The TSOFC is the fuel cell with the longest continuous development period, startizng in the late 1950s, several years before the AFC. The solid ceramic construction of the cell alleviates cell hardware corrosion problems characterized by the liquid electrolyte cells and has the advantage of being impervious to gas cross-over from one electrode to the other. The absence of liquid also eliminates the problem of electrolyte movement or flooding in the electrodes. The kinetics of the cell are fast and CO is a directly useable fuel as it is in the MCFC and ITSOFC. There is no requirement for CO_2 at the cathode as with the MCFC. At the temperature of presently operating TSOFCs (~1000°C), fuel can be reformed within the cell. The temperature of a TSOFC is significantly higher than that of the MCFC and ITSOFC. However, some of the rejected heat from a TSOFC is needed to preheat the incoming process air. The high temperature of the TSOFC has its drawbacks. There are thermal expansion mismatches among materials, and sealing between cells is difficult in the flat plate configurations. The high operating temperature places severe constraints on materials selection and results in difficult fabrication processes.

15.4 Cell Energy Balance

The energy balance around the fuel cell is based on the energy absorbing and releasing processes (e.g. power produced, reactions, heat loss) that occur in the cell. As a result, the energy balance varies for the different types of cells because of the differences in reactions that occur according to cell type. In general, the cell energy balance states that the enthalpy flow of the reactants entering the cell will equal the enthalpy flow of the products leaving the cell plus the sum of following terms:

(1) The net heat generated by physical and chemical processes within the cell,

(2) The DC power output from the cell, and

(3) The rate of heat loss from the cell to its surroundings. Component enthalpies are readily available on per mass basis from data. A typical energy balance calculation is the determination of the cell exit temperature knowing the reactant composition, the temperatures, H_2 and O_2 utilization, the expected power produced, and a percent heat loss.

15.5 Cell Efficiency

The thermal efficiency of an energy conversion device is defined as the amount of useful energy produced relative to the change in stored chemical energy (commonly referred to as thermal energy) that is released when a fuel is reacted with an oxidant.

$$n = \frac{\text{Useful Energy Produced}}{\Delta H \text{ (Change in stored Chemical Energy)}}$$

Hydrogen (a fuel) and oxygen (an oxidant) can exist in each other's presence at room temperature, but if heated to 580°C, they explode violently. The combustion reaction can be forced for gases lower than 580°C by providing a flame, such as in a heat engine. A catalyst and an electrolyte in the fuel cell, can increase the rate of reaction of H_2 and O_2 at temperatures lower than 580°C. It is also observed that a non-combustible reaction can occur in fuel cells at temperatures over 580°C because of controlled separation of the fuel and oxidant. The heat engine process is thermal where as the fuel cell process is electrochemical.

16

Techno-Economic Analysis of Energy Options

16.1 Introduction

It is well known fact that planners as well as end user's have to consider any resources as technology from economic point of view. Therefore, it is important to have techno - economic analysis of each energy source option. In the feasibility study of particular energy option, three tier judgement approaches is required, which include technical feasibility, economical viability and social acceptability. Hence in the rural energy planning programme one has to considered techno socio economic analysis of an individual energy option. Since in integrated rural energy planning, locally available energy resources are used, presuming that it is socially recognized. Hence in the IRPE project, the need and scope of techno economic analysis in terms of economic analysis of different technical feasible options, before between "economic" and "financial" analysis should be kept in mind. An economic analysis is needed to verify that a particular energy source will provide "net benefit" to society as a whole, before its final integration in the energy context, while a financial analysis integrated in the energy practices. In simple wards, a financial analysis is concerned with goods and services, money inflow and outflows and the transfer of payments. An economic analysis also considers the social benefits and costs of goods and services which are not traded in a market.

16.2 Frame Work for Economic Evaluation of Energy Practice

On economic evaluation of a particular energy source/practice should include the following sequential tasks-

1. A specification of the objective of the particular energy source considered, that is a reorganization tasks –
   (a) What the source is trying to achieve?
   (b) Who will be benefited from the source?
   (c) What problem is the source attempting to overcome?

2. An identification of the alternatives for achieving the specified objectives of a particular energy source. For each of the alternatives identified, the following steps are carried out: -
   (a) Identification and quantification of the physical inputs and outputs involved in the particular energy source.
   (b) Determination of the unit value in terms of both market prices and economic values, for the input and out.
   (c) Development of "value flow" tables, showing the total values of benefits and costs that are estimate to occur over the "useful life" of the particular energy source.
   (d) Calculation of the measure of particular energy source worth and other indices required to answer the questions of decision makers and user's.
   (e) Conduction of a sensitivity analysis in order to know the measures of energy source which might be altered with changed in the input – output relationship values.

16.2.1 Identification and quantification of physical input – output relationships

An evaluation of an energy source initially required the identification and quantification of the bio physical benefits resulting from the utilization of that source. Subsequently, economic evaluation attempt to "monetize" the quantitative and qualitative aspects of these benefits. Unfortunately, the task of quantifying these "input – output" linkage is difficult because.

(a) The bio physical relationship between a particular energy source and resulting benefit are not well known in many instance, since energy source in interrelation to ecology and environment.

(b) The information that links the source and benefit together in economic analysis often is not in a form that can be used readily.

One example to understand this approach is to consider different options for cooking energy requirement. Biogas is one of the options. It can easily identified and quantified physical input - output relationship for a particular size biogas plant. Say for 2 cum biogas plant, 50 kg dung plus 50 litre water as input and 2 cum biogas plus enriched manure as output. Further, one can think conventional utilization of 50 kg dung through cakes. However, an exact benefit derived from environmental point of view in quantification is a difficult task. The effects and benefit of a particular energy source are defined as changes or difference "with or without" the source being considered. In essence of biogas plant, when one says "increased energy availability" or "reduce environmental degradation", one should be referring to the difference in the energy availability and environmental degradation between conditions with and without the Biogas as energy source. Importantly, energy availability still can be declining or environmental degradation still cab increasing with a particular energy source, but at slower rates than without that source.

It is critical to keep the "with and without" concept in mind throughout an economical evaluation of a particular energy source. Ignoring this concept, result in significant error in measurement and analysis. The concept of "with and without is illustrating in Fig 16.1.

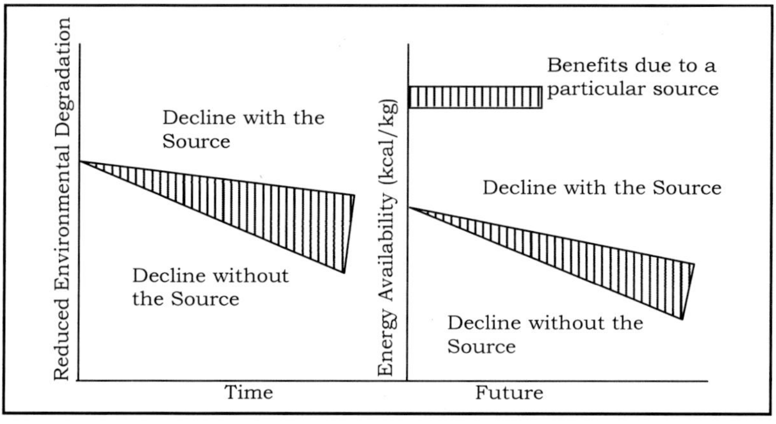

Fig 16.1. With and Without Concept for Economical Evaluation of Energy Sources.

(a) Identifying and Quantifying Input

The inputs in a particular energy source are those derived from the additional use of the factors of production, that is, land, labour, capital, equipment, raw material, structures and civil works etc. Input are considered to be divided between "on-site" and "off-site" effects. For analysis, the inputs to a Biogas plant can be categorized as follows:

1. *Land:* It depends on the size of plant.
2. *Labour:* It includes skilled and unskilled labour used during construction of a plant daily labour used.
3. *Equipment:* It includes the complete list of equipment required for construction, management and utilization of biogas and effluent slurry.
4. *Raw Material:* It is daily requirement plus water.
5. *Structures and Civil Works:* It consists of exact physical structure of biogas plant i.e. Input - outlet digestion assembly, pipe line lifting etc. in terms of main power and materials.

(b) Identifying and Quantifying Output

In many cases, it is necessary to conduct studies to "transform" input into output. Using the "with or without" approach, technical relationship, such as those presented in Table 16.1, can be developed for a particular energy source. In those technical relationships, the dependent variables are the output and the independent variables are the inputs.

Table 16.1 Technical Relationship in an Energy Source (Example Biogas Plant)

S.No.	Dependent variables	Independent variable
1.	Annual Direct Energy Availability (cum per in case of a biogas plant)	Change in annual pollution rates.
2.	Annual Environmental Degradation	Energy sources (Conventional dung use practice characteristics in terms of environmental loss)
3.	Annual Indirect Energy Availability (Such as enrich manure tonne per year)	Agricultural lands, Agricultural infrastructure (proportion of area)

16.2.2 *Determination of unit values*

The approach employed in valuing the benefits and cost associated with a particular energy source is straight forward. Market prices are used to value those inputs and output traded in the market place and non market benefits and costs are ignored. There are three approached to assign monetary values to the benefits and costs associated with an energy source.

1. *Market Price :* The market price itself can be used in the economic evaluation of an energy source as a reflection of economic value of the margin.
2. *Surrogate Market Price:* In the situation of benefits and costs not themselves valued in the market place, but for which "clear" substitute which are valued in the market exist, one can use appropriately adjusted market price for the substitute to develop surrogate values for benefits and cost being valued. Surrogate market approaches are often employed in deriving shadow prices, especially for unpriced "environmental benefits"
3. *Hypothetical Valuation :* In some cases, where there are no possibilities to derive acceptable market price measure of value, it is possible to drive value information through survey or through "expert judgment". However, because these approaches do not rely on market price, directly or indirectly, the results have to be interpreted carefully.

16.2.3 *Development of "value flow" chart*

A "Value flow" chart is the economic analysis equivalent of a "financial cash flow" table, is that it lists all of the benefits and all of the costs of a particular energy source on a year-by-year basis over the period of the analysis or the life of the entire energy source.

Considering value tables in general, following points need to be made:

1. Economic efficiency prices are used throughout. Sometimes, market price will be adequate measures of these economic efficiency prices (Energy Prices), in other cases, shadow prices will be employed or price determined by the use of other valuation method such as cost of environmental benefits.
2. All of the prices are in constant monetary limits. Inflation is not included.

3. All the on-site and off-site benefits and cost of a particular energy sources should be included in a value flow chart.

16.2.4. Calculation of the measures of energy source worth

Once the value flow chart has been prepared, it is necessary to evaluate the "stream" of benefits and cost to compare alternative energy sources. A benefit cost analysis is a systematic approach to make this comparison. In practice, net present worth, integrate rate of return and benefit cost ratio are three principal value measures which is utilized to know the project worth. However, annualized cost method or life cycle costing (LCC) method has also been preferred for economic evaluation owing to its simplicity and yet performing the task of comparing economic viability effectively. All these measure are calculate by using the some data making identical assumptions, with the exception that no discount rate is employed in calculation of the internal rate of return.

(a) Net Present Worth

This measure, also known as net present value (NPV) is based on the desire to determine the present value of net benefits from a particular energy source. If the goal of the economic analysis is to determine the total net benefits to society from a particular energy source, the net present worth (NPW) criteria provide a consistent ranking among the alternatives. The formula for the calculation of NPW is

$$NPW = \sum_{t=1}^{n} \frac{B_t - C_t}{(1+r)^t}$$

where

$B_t - C_t$ = Benefit or cost in year t

r = Discount rate

t = Time from year 1 to n

(b) Internal Rate of Return

The internal rate of return (IRR) is also frequently used to evaluate an energy source. Unlike the NPW or the benefit cost ratio, the IRR does not use a predetermined discount rate. Rather, the IRR determines the discount rate that sets the present value of the benefits equal to the present value of the costs. That is, the IRR is the discount rate r, such that:

$$\sum_{t=1}^{n} \frac{B_t - C_t}{(1+r)^t} = 0$$

The IRR utilizes the same information that is used in NPW calculation. Although the discount is not prescribed but determined as a result of the IRR calculation. This does not eliminate the use o prescribed discount rate. In an evaluation involving the IRR measure of worth, the calculation IRR is compared to some prescribed discount rate to decide whether or not an energy source is worthwhile.

(c) Benefit Cost Ratio

A third measure of worth is the benefit cost ratio (B/C ratio). This ration simply compares the present value of the benefit to the present value of the cost, as shown below:

$$B/C\ Ratio = \frac{\sum_{t=1}^{n} \frac{B_t}{(1+r)^t}}{\sum_{t=1}^{n} \frac{C_t}{(1+r)^t}}$$

If the B/C ratio is greater than 1, the present value of the benefits is greater than the present value of the costs. When this happens, the particular energy source meets the criteria of profitability. In fact NPV, IRR and B/C ratio (all these three measures) of an energy source worth are closely related. This fact is not surprising, since the same data on benefits and costs are used in calculating the three measures of worth. The values of these three measures of worth have the following general relations;

NPW	IRR	B/C Ratio
If >0	Than IRR> r	and >1
If <0	Than IRR<r	and <1
If =0	That IRR=r	and =1

Although all of the three measure of worth use the same data in their calculations and have a "symmetry" in their results, it is possible that, when a set of alternative energy sources are examined, different evaluation criteria can giving different ranking to the alternatives. This raises the question of which measure of worth NPW, IRR or the B/C ratio, to use. It is the NPW that is the "economic objectives" as it is this measure that seeks to maximize for the investment of available, but scare, resources. Importantly, NPW always must be a part of any "choice criteria' or ranking scheme for accepting or rejecting an energy sources. The IRR and the B/C ratio measures give no indication of the magnitude of the net benefits. Therefore, the reliance on either the IRR or the B/C ration can be lead to a result in which the total net benefits from an energy source selected are smaller that those benefits would be by using the NPW criterion.

(d) Annualized Cost Method or Life Cycle Costing (LCC)

It takes into account all costs including that of the investment spread over the entire useful life of the machine. It works on the principle that any new system should have the annualized cost (worked out from following equation) at least equal to that of an existing system to justify its replacement.

$$AC = (AC)_1 + AFC$$

$$(AC)_1 = CRF_1 + AOMR + AFC \qquad \text{.........(1)}$$

$$CRF = \frac{r}{1-(1+r)^{-n}} \qquad \text{.........(2)}$$

$$AMOR = F_1 \qquad \text{.........(3)}$$

Where

AC = Annualized cost of the system under consideration (Rs.)

$(AC)_1$ = Annualized cost component representable as a function of initial investment.

CRF = Capital recovery factor

AOMR = Annual cost of operation, maintenance and repairs (Rs/year)

AFC = Annual fuel cost (Rs/year)

r = Annual discount rate reflecting on the cost of capital advance to the user as per norms of funding agency

n = Useful life in years deduced from the expected yearly usage and total number of operating hours, the system could be run with ordinary repair and maintenance before a major replacement is necessary.

F_1 = Annual operating, maintenance and repair (AMOR) factor.

16.2.5 Conduction of a sensitivity analysis

Any time one becomes concerned with the future and one becomes concerned with uncertainty. Despite what any one might say it is not possible to predict the future, more the probabilities of different future. In general, uncertainty of an energy sources is associated with:

(a) Natural Factors

Especially those associated with natural climatic processes and their irregularities in occurrence such as solar energy and wind energy etc.

(b) Human Factors

Particularly those associated with utilization practices, management capabilities and impact of social, cultural and economic factors.

(c) Technological Factors

Including those associated with "structural" element in energy source.

Therefore, in an economic evaluation of an energy sources especially for Integrated Rural Energy Planning Programme, it is necessary to deal with uncertainty in the planning process. There are two interrelated steps involved in the consideration of uncertainty in the planning of a particular energy source.

(a) Identification of the likely sources of uncertainty. For each likely source, estimation of a "reasonable" rage of values that the parameter involved might take. Here, one can rely on the past experience and available evidence.

(b) Suggestion of appropriate way to re – design an energy source to reduce the chances of unacceptable results.

From a sensitivity analysis a planner can determine the magnitude of change in the measures of worth that is NPW, IRR, or the B/C ratio resulting from input - output relationship, value, size and scale use of a particular energy source.

16.3 Techno – Economic Energy Planning

The new and appropriate options or decentralized energy planning in the integrated Rural Energy Planning Programme could be considered in four major sectors as Cooking energy planning (domestic energy), lighting energy planning (Electricity generation strategy), Transport energy planning (Strategy for Electricity Generation).

16.3.1 Cooking energy planning

There are different sources available which can be used for meeting cooking energy requirement. Theses sources could be conventional as well as non conventional. However, it is essential to use energy source according to grade in cooking energy planning. This is not only required from energy conservation point of view, but also better quality of cooking food. In this context, it is important to know that every energy sources must be associated with a grade or quality, which may determined by its temperature or the temperature it can producer.

Such an approach along with optimum utilization of available resources based on techno-economic analysis for meeting out cooking energy requirement in rural area in given in Table 14.2. For cooking any item, the temperature requirement is approximately 100 -125 °C for certain duration depending on the nature of the material to be cooked. This can be fulfilled through cow dung cake, biogas, kerosene and electricity.

Table 16.2 revels that biogas is the optimum choice for meeting cooking energy requirement in rural areas. Further, biogas is not only limited for recycle of dung but in fact now, it is recognized as waste recycling and resource recovery system as waste material like biomass, forest waste, excreta, industrial waste and other municipal waste may also be anaerobically digested and may produce combustible gas. In addition to this, this approach may also eliminate the problem of pollution, which otherwise is being created on account of these wastes.

In rural areas biomass, agriculture and forest wastes are available in abundance. As such these materials are not efficiently used and are creating pollution therefore safe disposal of these materials is essential. Presently, in rural area mainly cattle dung is used for biogas generation.

However, night soil produces three times ore gas as compared to cattle dung. Mixing it with dung or with water hyacinth and algae could boost methane yield from 555 to 75 per cent. Further, rather than burning agricultural and forest wastes as ordinary fuel, it is better to anaerobically digest it for biogas generation or briquettes. It has been observed that feeding agricultural waste to the animal and the use the dung in biogas plants, gives an overall efficiency of 24 per cent, but charging these wastes directly into the biogas digester could give 60 per cent efficiency.

Table 16.2 Different Option for Cooking Energy in Rural Areas

S. No.	Alternatives Sources / Heating value in Indian context	Conversion Efficiencies (Percent)	Total Energy utilized (kcal)	Cost of useful 1000 kcal available	Other Constraint
1.	Cow dung Cake 2000 kcal /kg cost Rs. 1.50 /kg	10 Traditional chulha	200	Rs. 7.50	Low conversion efficiency, More deposit on Utensil, Effect on house wife's health
2.	Firewood 3600 kcal/kg Cost Rs. 2/kg	-do-	360	Rs. 8.00	Low conversion efficiency, Effect on house wife's health, Deforestation ad ecological imbalance
3.	Biogas 4700 kcal/cum cost Rs.5/cum	60 (Biogas burner)	2820	Rs 2.50	Better fuel for cooking, No effect on users health, Yield enrich manure, Convenient
4.	Kerosene 8900 kcal/lit Cost Rs. 11/lit	15 (Traditional Stove)	1339	Rs. 6.00	Inconvenient, Costly, Not available easily
5.	Electricity 860 kcal /kwh cost Rs. 4.5/kwh	70 (Hot Plate)	602	Rs. 7.50	Not suitable, Not easily available, Require more attention during cooking
6.	LPG (Button Stove) 12000 kcal/kg 301 per 14.5 kg	60 (LPG Burner)	8400	Rs. 2.50	Availability difficult. Dangerous in use
7.	Solar cooker	• Effectively can only be used for 200 days/year • Only one meal can be prepared • Completely in the hands of nature • Only about 500 gms of product can be cooked at a time in conventional one box type cooker			

Table 16.3 Option for Power Generation

S. No.	Alternative Options	Economics and other constraints under Indian context.
Centralized Power Generation		
1.	Coal based Thermal Plants (fossil fuel)	Rs 3.50 per kWh. This greatly depends upon the location and the transportation cost. Low conversion efficiency (41%)
2.	Diesel Power	Rs 4.25 Per kWh, high initial capital investment.
3.	Gas based Thermal Power Plant	Cost varing from Rs 4.00 to Rs.4.50 per kWh
4.	Nuclear Energy based Power Plant	Social constraints Non availability of raw material
5.	Hydro Power Generation	Capital intensive Site specific
Decentralized Power Generation		
1.	Biogas	Can replace about 100 percent petrol and 80 percent diesel in the operation. Require large quantity of biogas, about 0.65 cum gas per hour per kW electricity. Cost about Rs. 2.25 Per kWh
2.	Solar	Can be generated through thermal and photovoltaic route. High capital investment The cost is at least 2 to 5 times the price of power generation through wind farm
3.	Wind Energy	Aero-generators are site specific. Require sufficient torque or economically harnessed at 5 kmph speed. The cost of power from wing farms will be around Rs 3.25 to 4.00 in the first year depending on the machine and other factors. It can compute with the conventional power generation provided sufficient quantity of wind energy is available at the site.
4.	Wood based power generation	Technology is prefect and make use different type of biomass efficiently. Cost of power is around Rs. 2.75 Per kWh at the first year. The economics of co-generation power plant based on biomass seems to be much better. However, the capital cost the

16.3.2 Lighting energy planning

Electricity seems to be optimum choice for meeting out lighting energy requirement for small cottage industries, other production unit and other activities in rural areas. But centralized electricity generation is costly, besides high transmission and distribution losses. Decentralized electricity generation can be a cost effective power generation system based on renewable energy sources. Some option being investigated such as wind energy, biomass, biogas, solar thermal, solar photovoltaic, etc are being investigated. Among the various renewable options available today, wood based power generation through energy plantation is receiving increasing attention in the developing countries in view of its cost effectiveness and inherent storage capacity. Table 16.3 gives the comparison of different centralized and decentralized energy sources for power generation i.e. electricity generation.

However, biogas could also be considered for electricity generation. It could be directly used through gas mantles or be converted into electricity by using duel fuel gensets.

16.3.3 Agriculture energy planning

As far as Agriculture Energy Planning is concerned, it could be viewed through two angles. Firstly minimizing the dependence upon chemical fertilizer and other fossil fuel usage in the agricultural production. One of the best options for this purpose is to anaerobically digest waste material for biogas production. The gas production through this approach could be used for meeting out cooking energy requirement and for replacing petroleum for running engines for agricultural operations. In addition, this technology would yield enriched manure in the form of digested slurry, which could go a long way in reducing fertilizer consumption.

Second view of agricultural energy planning is utilization of efficient agricultural machinery. Here efficient indicate the energy efficiency of the machinery. The energy efficiency of agricultural machinery could be increased through utilization of alternative fuels. These alternate fuels could be based on new and renewable energy sources. Such as biogas, solar energy, biomass energy, etc. However, biomass could provide best option for replacing petroleum. Special energy crops like tapioca, sugarbeet, sweet sorghum, sweet potato etc. do not use large energy input, but could provide us power alcohol, which is one of efficient replacement for petroleum. Hydrocarbon plants like black quince and copaiba could yield diesel substitute that need to no refining.

References

Abbot, C.G. 'The silver disc pyrheliometer Smithsonian Misc. Coll, 1911, 56(19), 10.

Angrist, S.W. "Direct Energy Conversion", 2nd ed. (Boston, AA:Allyn and Bacon, Inc. 1971).

Anonymous, Gobar gas plant. IARI Publication, New Delhi.

Arnold, J. and J. Jongma, 'Wood as source of energy in developing countries; Preprint, FAO, Rome, 1977.

Backus, C.E. Ed. Solar Cells, Institute of Electrical and Electronics Engineers (IEEE) Press, New York, 1976.

Ball, R.S. Natural sources of power. Constable, London, 1908.

Bathe, G. Horizontal Windmills, Draft mills and similar air-flow engines. Philadelphia, 1948.

Batten, M.I. English Windmills, Vol. I. Society for the protection of ancient buildings, Architectural press, London, 1930. Vol. II by Smith, D., 1932.

Beckman, W.A. *et al.* "Design considerations for a 50 Watt photovoltaic power system using concentrated solar energy, Solar Energy 10(3) 1966.

Bickler, D.B. and E.N. Costoque. "Photovoltaic Cells and Arrays," in Record of the Photovoltaic Power Conditioning Workshop, Sandia Laboratories, Albuquerque, New Mexico (March 1977).

Boone, D.R. and M.P. Bryant, 1980. Appl. Environ, Microbiol', 40: 626-632.

Boopathy, R. 1987. Effect of temperature on anaerobic fermentation of coffee wastes. Asian Environ. 9(3): 13-18.

Bradley, W.J. "Designing and Sitting Solar Power Plants," Consulting Eng. 48(3) 80-84 (1977).

British Wind Energy Association, Wind Energy for the Eighties. Peter Perigrinus, Stevenage, 1982.

Brown, H. 'Energy in our future', in J.M. Hollander, and M.K. Simmons, Annual Review of Energy, 1-36(1)1976, Palo Alto, California.

Bruno, R. 'A correlation procedure for separating direct and diffuse insolation on a horizontal surface', Solar Energy, 97-100(20), 1978.

Bryant, M.P. 1979, Microbial methane production-theoretical aspects. J. Anim Sct. 193-201, 48.

Buch, F. A.L. Fahrenbruch and R.H. Bube. "Photovoltaic Properties of Five II-VI Heterojunctions," J. Appl.Phys. 48(4): 1977.

Buckiust, R.O. and R. King 'Diffuse solar radiation on a horizontal surface for a clear sky', Solar Energy, 503-09(21), 1978.

Buckland, S. Lee's patent Windmill. Wind & Watermill Section, Society for protection of Ancient Buildings, London, 1987.

California Wind Energy Forecasting System Development and Testing Phase 2: 12-Month Testing,EPRI, Palo Alto, CA, and the California Energy Commission, Sacramento,CA: 2003.

Calvert, N.G. Wind Power principles, Their Application on the small scale. C. Griffin & Co. Ltd., London, 1979.

Choudhury, N.K.D. 'Solar radiation at New Delhi' Solar Energy, 44-52, 1963, 7(2).

Collector performance', Transactions of the conference on the use of solar energy. Tucson. 1955, 3, Part II, 74-104.

Cooper, P.I. 'Some factors affecting the absorption of solar radiation in solar stills,' Solar Energy, 373-81(13), 1972.

Coulson, K.L. 'Solar and terrestrial radiation methods and measurements', Academic press, New York, 1975.

Cuomo, J.J. and J.F. Zieflar and J.M. Woodale, 'A new concept for solar energy thermal conversion', Appl. Phys. Lett., 557(26), 1975.

Currin, C.G. *et al.* "Feasibility of Low Cost Silicon Solar Cells", Solar Cells (New York: IEEE Press, 1976).

Dave, J.V. 'Validity on the isotropic distribution approximation in solar energy estimation', Solar Energy, 331-333(19), 1977.

de Little, R.J. The Windmill. Yesterday and Today. J. Baker, London, 1972.

Dermatis, S.N. and J.W. Faust, Jr. "Semiconductor sheets for the manufacture of Semiconductor devices", IEEE Transactions Communications Electronics (New York: IEEE Press, 1963).

Dhirendra, R.P.S. Srivastava, and D.R. Karnure, 1986. Energy from solid waste, Biogas digester based on cattle dung-night soil mixed feed proc. BESI. 3rd Convention and Symposium, 30th Nov-2nd Dec. 1986. Pune.

Dohne, E. 1980. Biogas storage and utilisation In: D.A. Stafford, BI Whealtley and DE Hughes (Editors) Anaerobic Digestion. Applied Science Publishers Ltd. London pp 429-448.

Drummeter, L.F., Jr. G. Hass, 'Solar absorptance and thermal emittance of evaporated coatings', Physics of thin films, Academic Press. New York, 305-361(2), 1964.

Drummond, A.J. 'The extraterrestrial solar spectrum.

Duffie, J.A. and W.A. Beckman, 'Solar energy thermal processes', Wiley - Interscience publication, 1974.

Eckert, E.R.G. and R.M. Drake. 'Analysis of heat and mass transfer, Mc-Grawhill, Inc., 1972, p. 586.

Eckholm, E. 'The other energy crisis; firewood' World Watch Paper 1. World Watch Institute, Washington, 1975.

Fair, G.M. and E.W. Moore, 1937. Observations on the digestion on sewage sludge over a wide range of temperature, sew. Works J. 9:3.

Fowler and Joshi, 1923: Referred in Garg *et al.* 1980. Designing a suitable biogas plant for India. Appropriate Tech. 7(1).

Garcia, J.L. 1982. Advances in digestion anaerobia. Symposia International, Mexico, 25-27, October 1982.

Garg, H.P. and C.L. Gupta, 'Design data for direct solar utilization devices, Part II - Solar radiation data', Indian J., Met. Geop., 221-226 [20(8)], 1980.

Garg, H.P. and C.L. Gupta, 'Design data for direct solar utilization devices - 1: System data', J. Inst, Engrs. (India). 461-470[47(a)], 1968.

Garg, H.P. and P.C. Pande, and K.P. Thanvi, 1980. Designing a suitable biogas plant for India, Appropriate Technology. 7(1):29-31.

Garg, H.P. Ram Chandra and Usha Rani, 'Transient analysis of solar air heaters using finite difference techniques', Energy Research, 243-252(5), 1981.

Goldemberg, J. 'Global operations for short range alternative energy strategies', Energy, 733-744(4), 1979.

Government of India, 'Report of the working group on energy policy', planning Commission, New Delhi, 1979.

Hammond, A.L. "Photovolatics: The semiconductor revolution comes to solar", Science. 197 (4302): 445-447.

Hartnett, J.P. 'The case of alternative energy sources: Chapter 2, 19-38, from 'Alternative Energy Sources, Ed. J.P. Hartnett, Academic Press, 1976.

Harverson, M. Persian Windmills. The International Molinological Society, Reading, 1991.

Haurwitz, G. 'Insolation in relation to cloud type', J Meteor, 110-113(5), 1948.

Hickok, F. "Handbook of solar and wind energy".

In Situ Performance Measurements of the MITRE Photovoltaic Array (Washington, DC: The MITRE Corporation, 1976).

IREDA News Vol II, No. 3, July-September 2001.

Kalnay, E., S.J. Lord and R.D. McPherson, 1998: Maturity of Operational Numerical Weather.

Khandelwal, K.C. and S. Mahdi. 1986. Biogas Technology-A practical handbook, Tata Mc-Grawhill Publishing Co. Ltd. New Delhi.

Klein, S.A. 'Calculation of flat plate collector loss coefficients', Solar Energy, 79-80(17), 1975.

Klein, S.A. J.A. Duffie and W.A. Backman.'Transient considerations of flat-plate solar collectors', J. Engineering for power, Trans ASME, 109-113(96A), 1974.

Klein, W.H., B. Goldberg and W. Shropslire, Jr. 'Instrumentation for the measurement of the variation, quantity and quality of sun and sky radiation', Solar Energy, 115-122 [19(2)], 1975.

Kondratyev, K. Ya 'Radiation in the atmosphere', Academic Press, New York, 1969.

Kondratyev, K. Ya, and G.A. Nikolsky, 'Solar radiation and solar activity' Quart, J. Roy, Meteorol Soc., 509-522(96), 1970.

Kondratyev, K. Ya, 'Radiation processes in the atmosphere' World Meteorological Organisation, Geneva, WMO No. 309.

Kreith, F. and J.F. Kreider, 'Principles of solar engineering', Hemisphere Publishing Corporation, Mc Graw-Hill Book Company, 1978.

Kristofer Son, K.L. and Bokaldess, V., Renewable Energy Technologies. Their applications in developing countries. Beijer Institute. The Royal Swedish Academy of Sciences, Pergamon Press, New York, 1986.

Lampert, C.M. 'The use of coating for enhanced solar thermal energy collection', Lawrence Berkeley Laboratory, Report No. LBL. 8072, April, 1979.

Liu, B.Y.H. and R.C. Jordan, 'Daily insolation on surfaces tilted towards the equator', ASRAE Journal, 53-59 [3(18)], 1961.

Majumdar, N.C., B.L. Mathur and S.B. Kaushik, 'Prediction of direct solar radiation for low atmospheric turbidity', Solar Energy, 383-394, [13(4)],1972.

Makhijani, A. and A Poole 'Energy and agriculture in the third world, Ballinger Press, New York, 1975.

Marty, B. and S.A. Germe, 1984. Microbiology of anaerobic digestion. In: AM Bruce, A Kouzeti-Katsiri and PJ, New-man (Editors) Anaerobic Digestion of Sewage Sludge and Organic Agricultural Wastes, Elsevier Applied Science Publishers, London/New York: pp 72-87.

Mathur, A.N. and N.S. Rathore, 1992. Forestry for arid regions, Agrotech Publicatios, Udaipur.

Mathur, A.N. and N.S. Rathore, 1996. New and renewable energy sources. Bohra Ganesh Publications, Udaipur.

Mattox, D.M. 'Solar energy materials preparation techniques', J. Vac. Sci. Technol. 1023-1031, 12(5),1975.

McGuigan, D. Small Scale Wind Power. Prism Press, Dorchester, 1978.

MITRE Photovoltaic Energy Systems (Washington, DC: The MITRE Corporation, 1975).

Moulik, T.K. and Shukla P.R., Integrated Rural Energy System - Planning & Organisation, Indian Institute of Management, Ahmedabad, 1987

Mubayi, V. and P.M. Meier, Energy models for developing countries. A comphrenshive Assessment, National Centre for Analysis of Energy System, Brookhavan National Laboratory Upton, New York, 1973.

National Productivity Council, Productivity Promotion in fuel & heat utilization, NPC, APO study mission report, New Delhi, 1979.

Orgill, J.F. and K.G.T. Hollands, 'Correlation equation for hourly diffuse radiation on a horizontal surface', Solar Energy, 357-359(19), 1977.

Parikh, J.K. 'Energy systems and developments', Oxford University Press, Delhi, 1980.

Peterson, R.E. and J.W., Ramsay, 'Thin film coating in solar thermal power systems', J. Vac. Sci. Technol., 471(12), 1975.

Photovoltaic energy program summary, "Solar Energy Programme Anthology" (Washington, DC: Energy Research and Development Administration (ERDA), 1971.

Prediction: Medium range, Bulletin of the American Meteorological Society, 2753-2769(79).

Rao, K.R. and T.N. Seshadri, 'Solar insolation curves' Ind. J. Met. & Geop., 1961, 12(2), 267-272.

Rathore, N.S. Panwar N.L. and Surendra Kothari ' Biomass production and Utilization Technology, 2007, Himanshu Publication, Udaipur, Rajasthan India.

Rathore, N.S., A.K. Kurchania and N.L. Panwar 'Non Conventional Energy Sources', 2007, Himanshu Publication, Udaipur, Rajasthan Inida

Robinson, A.L. "Amorphous Silicon: A new direction for semi-conductors", Science 197 (4306): 851-853 (1977).

Ruth D.W. and R.E. Chant, "The relationship of diffuse radiation to total radiation in Canada", Solar Energy, 153-154(18), 1976.

Sathianathan, M.A. 1975. Biogas achievements and challenges. AVARD New Delhi. Singh, RB and Singh, KK 1978. Janta biogas plants preliminary report on design and cost. PRAD, Lucknow.

Siddharatha Bhatt, M. The Design of Rural Energy System, Proc. Indian Acad. Science (Engg. Sci.) 7(3), Nov. 1984.

Smith, E.V.P. and D.M. Gottlieb, Space Science Rev., 771(16), 1974.

Stair, R. and H.T. Ellis, 'The solar constant based on new spectral irradiance data from 3100 to 5300 Augstroms 'J.Ap. Meteorol, 635-644(7), 1968.

Steinhart, C.E. and J.S. Steinhart, 'Energy: Sources use and role in human affairs', North Scituate, Mass., Dubary Press, 362, 1974.

Stout, B.A. 'Energy for world agriculture, FAO Agriculture Series, No. 7, FAO, Rome, 1979.

Survey of Energy Resources, 1974. US National Committee of the World Energy Conference, New York. 1974.

Tabor, H. 'Selective surfaces for solar collectors', Low temperature Engineering Application of Solar Energy, ASHRAE, New York, 1967, Chapter 4,41-52.

Taylor, T.B. 'A preliminary assessment of solar energy technology for world wide use': Report to the Rock feller Foundation. 1979.

Thekackara, M.P. 'Data on incident solar energy', Supplement to the Proceedings of the 20th Annual Meeting of the Institute for Environmental Science, 1974, 21.

Thekaekara, M.P. 'Solar radiation measurement techniques and instrumentation', Solar Energy, 309-325, [18(4)], 1975.

Thekaekara, M.P. 'Solar radiation measurement: techniques and instrumentation', Solar Energy, 309-325(18), 1976.

Threlkeld, J.L. and R.C. Jordan, 'Direct solar radiation available on clear days', Am. J. Heating Piping air Conditioning, 135-145(29), 1957.

Twidell, J.W. and A.D. Weir, Renewable Energy Resources. University Press, Cambridge, U.K., 1986.

Van Dijk, H.J. and PD Goedhart, Windpumps for Irrigation, C.W.D., Amsterdam, 1990.

Vidyarthy, G.S. 1977. Prospect and achievements of biogas programme. AET 1 (ix): 2-4.

Vimal, O.P. and M.S. Bhatt, Wood Energy System, K.L. Publications, New Delhi 1989.

Warman, H.R. 'Future problems in petroleum exploration' Petroleum Review', March, 1971.

Weinberger, H. 'The physics of the solar pond', Solar Energy, 45, [8(2)], 1964.

Westrick, K., 2003: Description and results from a comprehensive regional wind energy forecast system in the Pacific Northwestern U.S. *Proceedings of Windpower 2003 Conf.*, Austin, TX, American Wind Energy Assoc.

Whillier, A. 'Design factors influencing solar collectors in Low temperature engineering applications of solar energy', ASHRAE, New York, 1967.

Whillier, A. 'Solar energy collection and its utilization for house heating, ScD. Thesis, MIT. 1953.

Wiebelt, J.A. and J.B. Henderson, 'Selected ordinates for total solar radiation property evaluation from spectral data', Trans. Am. Soc. Mech. Engg., J. Heat Transfer, 101, 1979.

Wolf, H.C. Efficient use of Energy. AIP conf. Proc. No. 25, American Institute of Physics, New York 1985.

Wolff, A.B. The Windmill as Prime Mover, 2nd edn. Wiley, New York, 1885.

Wood, B.D. 'Solar energy measuring instruments', Solar Energy Engineering, Academic Press, London, 1977.